Annual of
LANDSCAPE DESIGN
of China

中国景观设计年刊

2008

ANNUAL

I

中国建筑设计研究院环境艺术设计研究院
中国风景园林学会规划设计专业委员会
世界建筑导报社
主办

北京筑语图书工作室
主编

华中科技大学出版社

图书在版编目（CIP）数据

中国景观设计年刊. 2008 / 北京筑语图书工作室编.
武汉：华中科技大学出版社，2008.7
ISBN 978-7-5609-4677-1

I. 中… II .北… III .景观－园林设计－中国－年刊
IV.TU986.2-54

中国版本图书馆CIP数据核字（2008）第095379号

中国景观设计年刊. 2008
出版发行：华中科技大学出版社
地　　址：武汉市珞喻路1037号（邮编:430074）
出 版 人：阮海洪
责任编辑：张亦男
责任监印：张正林
排版印刷：北京方嘉彩色印刷有限公司
开　　本：965mm × 1270mm 1/16
印　　张：19.5
字　　数：156千字
版　　次：2008年7月第1版
印　　次：2008年7月第1次印刷
ISBN 978-7-5609-4677-1/ TU · 360
定　　价：299.00元（含光盘1张）

联合出版：天津市天水文化传播有限公司
邮购电话：022-27461713
地　　址：天津市南开区鞍山西道291号

Annual of LANDSCAPE DESIGN of China

中国景观设计年刊

2008

ANNUAL

I

Preface
序言

WU Jiahua 吴家骅

What is scenery?
景观是什么

ANNUAL
I

在最近的电视节目里，我发现有个自以为是的主持人邀了一帮子老实巴交的所谓学者和热心人争相讨论要不要重建圆明园。其间，有嚷着要环保的，其实就是学种树的那种；有捶胸顿足嚷着要重建的，其实就是所谓文化人的那种。这帮人也真是老实，全体被那个电视台的小子给捉弄了。你们吵什么吵？事情其实没有那么复杂。

景观是个又好看，又有用的东西，所谓“pleasure with profit”讲的就是景观设计的根本。国土资源如此紧缺，中看不中用的事情不能干。给后人留点余地，留点老本才对。

我以为：景观可以是麦田、是果树、是鱼池，可看、可食；景观可以是地道、停车场、马路牙子，可交通；景观可以是步行场所、楼前房后，可交易、可谋生。也就是说，景观除了赏心悦目之外还给人以实惠。

如果不会做，就慢慢学，千万别动不动就用环保、文化、民族精神之类自己吓自己，忽悠别人或被人忽悠。如今这个年代，真理好像都掌握在媒体主持人手里，怪事！

I saw in a recent TV program a smug presenter hosting a discussion on whether or not to reconstruct the Summer Palace, and a group of so-called scholars and enthusiasts were among the participants. Some of them mentioned environmental protection, meaning as many trees should be planted inside as possible; some others, who label themselves as cultured people , strongly advocated a full copy of the destroyed garden. How stupid you guys are. All of you have been smeared by the host. What are you people quarrelling for? In fact things have not been as complex as you assume.

Scenery should be something of both beauty and helpfulness. Someone puts it like "pleasure with profit", emphasizing the fundamentals of scenery design. Domestic resources are so scarce that we should not allow any space for those more pleasant to the eye than use. We should leave our posterity some spare land, some capital.

I think scenery could be the cornfield, fruit tree or fish pond, which are reserved for watching or food. It could be a tunnel, parking lot or roadside stone built for traffic convenience . It could also be a footpath and space around people's home that could be traded or used for living one's life.

If you do not know how to do it, relax and learn step by step. Never be intimidated by the concepts of environmental protection, culture or nationality spirit, or fool others into being so cowed. How weird it is that in these days the hosts appear to have mastered the truth.

Contents
目录

City Park 城市公园

City Design 城市设计

Contents
目录

Scenic Spot 旅游景区

Residential Community 居住社区

Annual of Landscape Design of China is presenting its third book. Every time before the publication, the editorial would ask us, a sponsor of the book, to say something. But we used to reject the invitation, on the grounds that the contents are most critical and a magazine needs to do more than saying. This time, however, we would not deny the invitation, because the matured industry makes us feel necessary to say something about the yearbook, the design and the sector.

THE YEARBOOK

One day in early 2004, Mr. Rao Liangxiu, editor-in-chief of Yearbook of Chinese Interior Design, came to me with a proposal concerning editing Annual of Landscape Design of China. This was good news for me who has been involved in the design cases over the years. My experiences have taught me that an individual or a team can hardly master the industry trend, that the prosperous segment demands all the job takers concentrate on it, and that a developed segment can help position and orient the people to further benefit the local scenery design sector. All these things need a platform and a yearbook may do something for the segment. Our work started therefore.

The starting book received support from the Chinese Society of Landscape Architecture that appointed the Environment Arts and Design Research Institute of China Architecture Design & Research Group and the Planning Committee of the Society as sponsors of the book. Ms Yu Bing, who is now editor-in-chief of Domus (Chinese version) and then editor-in-chief of Beijing Editorial of the World Architecture Review, took part in too helping the idea come true. A group of experts were invited and a committee of them was established, headed by Mr. Cui Kai, chief architect of the Institute, together with an editing committee and an editorial, which is responsible for the routine job of the book. The first book came in spring of 2005 and 2006 saw the second the now the third.

Handling of the yearbook has been a tough job and its survival needs to thank support from various sides. An article soliciting letter clarified the objective: to create a platform for local designers and do something for the sector. Our efforts have been paid off. The book has received support from peers and the designers as well. There would not have been so many high quality articles without the support.

The book needs further improvement, better articles and greater characteristics of course. We hope that in the near future, the book would have become one reflecting the industry level, the achievement and the professional spirit, one expected and cherished by everyone.

THE DESIGN

I am not talking about how to do the design. I mean how to finish the job, which can reflect a person or a team's industry spirit to some extent. It is in fact what I said before to do something for the benefit of the segment.

First of all, the design is not individual. The designers need to bear in minds the entire industry. There was a seminar on how to maintain the design charge level, which is a problem of the industry. On a market where the situation is yet so good, a lower than average offer from a designer is going to affect the evaluation of a consigner. And a perfect design needs to be better paid for the sustainability. We have been sticking to the point to win the trust of consigners. Higher pay guarantees better quality, this becomes acceptable for the consigners.

Secondly, a good design is innovative. A designer needs to create perfect product. The design should not be done for the sake of profit, or the approval from consigners. A good designer needs to have long-range vision, instead of copy and imitation. Higher level requirement means positive manner in the job. Only most designers do their jobs in this way, can the industry be developed, be in line with the world, and local masters be fostered. Over the years, we have been doing our jobs like this in the case as following: the Bird Nest, the Potala Palace and the Zongjiao Lukang Park, Xidan Shopping Street, Xichang Torch Festival Plaza, Lhasa Train Plaza and couples of scenery designs for Vanke Company. It has been approved that our ideas are accepted by every consigners and we are confident to carry out them to the end.

What is more, the design should be eco-friendly. A designer needs to bear in mind a social responsibility sense. Ecology is not a slogan or concept, it is an attitude towards design and the life. If we are willing and put our efforts on it, it will reflect our ideas and affect our life. In our designs, we have repeatedly maintained the existing trees, rivers and landforms; we have tried our best to adopt more water penetrating materials to give way to rainwater balance; we have treated the scenery plants just like with our bodies and we have fully considered the balance between the design and the eco environment after it. What we mean is that an eco sense is good for the industry progress and social development.

THE INDUSTRY

What is the job we are doing now? There are confused interpretations now. I think it reflects an immature industry, one compared with other segments. It also means the industry allows for opulent imagination. Therefore, I would like to present two proposals: whatever happens, we need to keep the dispute a scholar matter. The second suggestion is whatever one deserves, he or she needs to do things in real sense.

Perhaps being motivated by such ideas, we have created such a book, which has an open mind towards various sides. What is more, our publications include the Scenery Design magazine and many others. In doing these work, we have learn something, won confidence, made more friends and developed ourselves.

On the occasion of this book, may I wish a better publication, a prosperous industry, more participants in doing the same things for the industry.

KONG Xiangwei 孔祥伟

Mentioned Environmental Protection
景观设计评论家
北京观筑景观设计有限公司首席设计师

Who are we?

A recognition and reflection on modern scenery designers

我们是谁？

——当代景观设计师的认知与反思

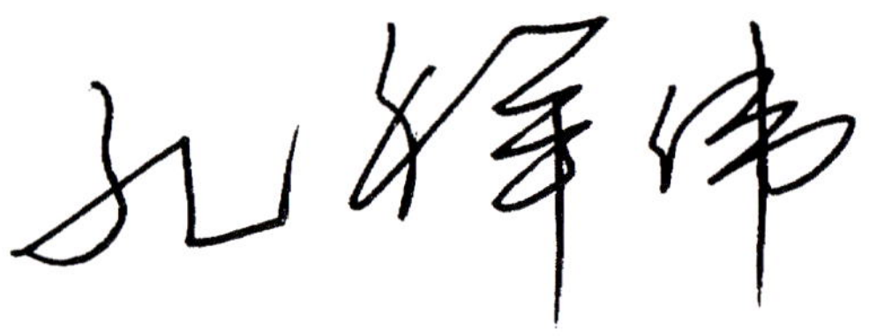

"我是谁？" 这是莎士比亚笔下哈姆雷特的追问，这是一个有关自我认知的自省式问题。景观设计师是谁？在当下的设计环境中，这是一个所有景观设计师都要自我思考的问题。

自20世纪90年代以来，中国快速城市化运动改变了中国的步伐，同时将中国的大地变为一个大的工地，这催生了许多新兴的行业。一个称之为"景观设计"的行业走到人们的视线中来。"景观设计"这个词汇携带了太多的信息，带着争议、带着沉重、带着希望、带着美好的想象，甚至带着娱乐式的美化功能被人们所使用。而景观设计作为一个新兴职业，某种程度上也成了一种玩票的时髦的游戏式的职业，随意描摹，随意在大地上留下划痕和印记。

现状与反思

中国普遍的景观是一种怎样的现状？在非设计领域，有致命性的破坏（诸如国土资源的山体破坏、水体污染、生态系统割裂），物种消失和错误性的建设（诸如大地的随意覆盖、宽马路、大广场、城市建筑间隙的硬化到城市河流的渠化等等）；在设计领域，景观设计主要是一项改善人居环境的工作。全国各地人居环境范例城市的评选、房地产领域人居环境奖的设立，为这项工作提供了用武之地。景观设计师单方面从人的角度出发进行设计，而这里的人绝大多数是某些个体和小的群体，譬如单个决策者和利益群体，并没有将公众纳入其中；再者，便是对自然和生态环境的漠视，杀鸡取卵，将自然中的资源转变为小区域中的利益群体所有。不可否认，中国在快速城市化发展的进程中，出现了许多所谓的"高品质"的城市公共空间和"高品质"的居住空间环境，愉悦着人们的身体和视觉，但对视觉美却缺乏辨析，将一味的视觉美等同于设计真谛，却丝毫预料不到美艳背后的危机。

咎其原因，错综复杂。首先是概念的混淆与现代设计理念的缺失。在概念上将景观设计等同于绿化设计，等同于表象的美化环境的设计；而现代理念的缺失是造成中国景观设计现状的直接原因。对景观设计而言，现代的设计理念意味着强烈的时代性，意味着积极应对当代的系列危机问题，而不是去回避，而不是打着弘扬历史的旗号，运用某个时代的历史语言，或者运用西方现代设计早已摒弃的西方古典语言——所谓的欧陆风格、法式、意式、英式风情等等；现代设计理念来源于与我们生活密切相关的现代艺术，来源于延续几千年仍被生产和生活使用的此时此地的乡土语言；现代设计理念尊重土地和自然，尊重大众和平民，尊重生态和乡土，尊重可持续设计理念，尊重全球化并尊重地域差别，同样尊重草根文化；现代设计理念不放弃现代主义的优点，同样不离弃后现代主义的多元化理论；现代设计理念不是纯粹的生态学法则的再现，它承认"每一件事情都与别的事情有关"，但不因"人类所做的对地球都是有害的"警告而止步不前；现代设计理念认同可持续发展的观念，致力于在自然和人类并存的前提下，或者说缝隙中使二者之间的关系走向融洽；现代设计理念将生态学的思想作为一切行为的前提，作为检验设计工作的标尺。

其次是设计行为作为商品交换的环节和公众参与的缺失。在当代，设计无疑也是商业活动的一种方式，设计师的作品，在绝大多数的时候是商品，取决于买方市场的态度，或者说是消费者的态度。甲方个人的喜好决定着一个设计成果，而这些设计成果对于我们脚下的土地是一次性不可逆的行为，俞孔坚甚至将我们生活的城市和赖以生存的自然现状，称之为"五千年未有之破坏。" 庞伟这样批判到："这是一个犬儒主义盛行的时代，只认外在的世俗功利，否认内在的德性与价值"。王尔德说："犬儒主义者对各种事物的价钱一清二楚，但是对他们的价值一无所知"。 庞伟将这种现象称为"富饶的贫困"。 中国每年消耗世界水泥产量的40%，而每天新增的建筑面积占到全球总量的近半，在这些惊人的数字背后的中国建筑师是怎样的一群人？中国今天的建筑繁荣与世界建筑史有怎样的关系？我们在建筑什么？为什么建筑？"这是一场富饶的贫困，纷繁热闹的表象之下，是精神出口的狭隘、意义空间的遗丧。"

社会学者李津奎曾经这样讲，一个刚踏上工作岗位的城市规划师，在一个布满等高线、山地、湿地，河流水塘的自然之上，

顷刻间分割出"系列方格子"，图面上的细线所到之处，将自然肌理漠然切割，同时将拥有几百年甚至上千年的村落掩埋。这绝不是耸人听闻之言，这样的例子比比皆是，我们所生活的城市就是这样形成的。

将庞伟笔下的建筑和李津奎话中的城市置换为景观，同样如此。设计师的随意一笔都将在大地上留下永久性的划痕，近十年来，中国的设计师将中国大量的土地覆盖，阻隔了土地与空气之间的自由呼吸。

出于利益的交换，与土地相依存的农民将世代敬畏而留存下来的护佑家园的大树亲手挖掉运往城市；出于利益的交换，设计师极尽想象之能事，以视觉美艳的设计语言和震撼人心的巨构赢得甲方的好感。中间有几许被动？几许主动？已不好言说。

还有就是审美趣味的混乱与价值观的错位。当代大众眼中的美的事物，带有强烈的现实主义的功利性。中国人在对拥有精神寄托或文化寄托的事物上的认识开始变得扭曲：风水、古代意境、符号这些事物便随之而来，同时叠加着拙劣的巴洛克符号，贴着所谓的"西古典主义"风格，或者说各种风格都不算的所谓的奢华风格！回顾现代化主义运动初期，西方设计师，以路斯为首的对传统的批判——"装饰即罪恶"之后，评论家称西方设计师从爱好古董的情结中解脱出来。而当代中国人似乎是天生的"后现代主义者"。

审美趣味的混乱和价值观错位直接决定了我们的城市景观对自然构成的巨大的影响。中国当下流行的景观审美，一是中国古典主义——以自然中的与水土保持和生态系统息息相关的"奇石"和大树为主要元素的造景方式，意味着对自然的掠夺；二是西方古典主义——以精致的工艺见长，布置修剪整齐的花木、动态的喷泉，铺砌光亮、奢华的大理石庭院，耗费着巨大的维护成本和资源，极尽繁琐与奢华，这都意味着与可持续发展的对抗。

而我们这个时代需求的审美趣味应以生态为基础，兼顾一切事物的审美趣味，而非视觉的美艳。美来自于平衡与和谐，时代呼吁乡土之美、野草之美，甚至于廖荒之美和遗产的锈蚀之美。概括一下，即可持续之美亦或伦理化之美。

景观设计师是谁

而景观设计师是谁？景观设计师是始终踏在土地上的人，俞孔坚将之概括为"处理土地与人的关系"的人，认为景观设计师不是鲁班的后人，而是大禹的后人，要么陶醉于小桥流水和风景的"艺术"，要么勇敢地承担起设计和协调中国人与生态关系的重任，再造安全、健康的秀美山川。

在与庞伟的对话中，他提到，我觉得在今天，人类共同遵守的东西其实不多，但一定要有这些东西。比如今天对于种族之间的攻击性话语一定是不能进入文明语言的，两性之间攻击性的话语也是不能进入文明语言的，同样，保护生态的问题，也是文化的基本点。这些方面一定要做到正确。还是援引庞伟的话："设计师的行为已然成为市场规律下的惯性行为，在这个唯利是图的时代，我们还怀疑什么？我们还有思考吗？我们还存在吗？"

有着共同的守护与约定，景观设计师才具有集体称谓的意义。景观设计师缺乏对生态系统的计算，一栋建筑，如果不经过严格的计算，建筑设计师的方案再美丽都会轰然坍塌，而景观设计，面对土地，我们却没有计算，如果始终没有计算，我们土地的承载力同样是有限的，只不过反应速度慢而已，终有一天，大地也会坍塌，而坍塌的力量会带来难以想象的毁灭。

我们往往漠视这些反应，景观设计师是谁？景观设计师是一个能预知事物反应的人，是与一个与生命体站在一起的人，而不仅仅是一个静态物体的设计者。

在文章最后，援引一位美国景观设计师的话："我不认为我们总能够制造迷人的景观，我认为我们应该创造深层次的、意义深远的景观。可以说建一个建筑就像一夜情：非常的热辣、非常的性感，但那只是一夜情；一个好的景观应该像婚姻：那是深层的、意义深远但是永远难以达到巅峰的体验。"

"Who I am?" This is a question from Hamlet in Shakespeare play. This is a question for self recognition. What is a scenery designer? This is a question for all the designers in nowadays China.

The fast-going urbanization campaign has fastened China's footstep ever since the 1990's and China is now a large construction site. The campaign has given birth to a lot of new sectors and scenery design is among them. The term scenery design is too informative. It means dispute, weightiness, hope, positive imagination and recreation to some extend. And the fashionable occupation like scenery design means the people can make marks on the land freely.

CURRENT SITUATION AND REFLECTION

How are the ordinary sceneries? There have been fundamental damages, such as the broken mountain, water pollution, disconnection of the eco system, species extinction as well as the wrong constructions, including the covered land, wide street, big square, cemented space among the city buildings and artificial canals of city rivers.

The scenery design work is expected to improve the human environment. But some designers and collectives have failed to fully consider the interests of the public. The other bad thing is to neglect the natural and eco environment, borrowing the natural resources to the use of a smaller collective.

It is no deny that there have been a lot of "high caliber" city and living space, making people feel comfortable in their bodies and vision. But so many people have failed to tell the difference between the visional enjoyment and the crux of the design, being unaware of the crisis behind the beauty.

The reasons are complicated. First of all, it is because of a lack of the modern design philosophy. The scenery design is assumed equal to greening design, to the design to beauty the environment. And a lack of the modern philosophy is the immediate reason for the current situation.

For scenery design, the modern philosophy means an active response to the series crisis, rather than any avoidance, rather than, under the banner of carrying forward the history, adopting the historical code of a specific time, or applying the dead ancient tactics-the so-called Europe, French, Italian or British touches.

The modern design philosophy refers to the modern art closely linked to our life, the local language that has survived over the past thousands of years and still being used. The modern philosophy respects the land and the nature, the public and ordinary people, the ecology and local land, the sustainability, the globalization and regional difference as well as the grass root culture. The modern philosophy does not abandon the advantages of modernism, nor the multi-end theory of it.

The modern philosophy does not copy pure ecology rules. It admits every thing is linked to others, but does not limit itself because of the warning that everything people do is harmful to the globe. It echoes the sustainable development concept, and, under the condition of a co-existence of the nature and human being, tries to harmonize the relationship of the two. It puts the ecology theory as a top priority of any actions, a staff guage to prove the design job.

The other bad thing of the purchased design job is a lack of the public participation. In our time, design is no doubt a kind of commercial activities. In most cases, the designers do their job for sale, which is determined by the consumers. The determination is a one-off and irreversible event for the land under our feet. Mr. Yu Kongjian calls the current situation "an unprecedented damage over the past 5,000 years".

China consumes up to 40 percent of the world's total cement production annually and nearly half of the newly added buildings stand in China every day. How are the architects behind such astonishing figures. What is the relationship between China's prosperous construction market and the world architect history? What are we building and why?

A sociologist Mr. Li Jinkui tells us such a story. A newly employed designer would instantly draw the panes on a draw full of contour lines, hills, wetlands, rivers and ponds. The panes cut the natural texture calmly and bury a village with a history of hundreds or even thousands years. This is not an exaggeration. There have been a lot of such cases. Our cities have been made in this way.

Over the recent 10 years, the local designers have cemented a lot of land, cutting the free communications between the land and air. For the sake of money, villagers replant ancient trees in cities from their yards; for the same reason, the designers would try to attract the potential purchasers of house with visional languages and charming structures. Who can tell what are behind all these things?

The following reason is the taste disorder and the mismatch of value concept. The beautiful things in the public eyes are based on the strong utility of them and there have been twisted understanding over things that carry spirit or cultural meanings. Fengshui, ancient touch and symbol, all these things are finding use, on top of the Baroque symbol, the so-called new classics and the luxury style that can not be classified. Chinese in these days are just like born post-modernists.

As a result, the fashionable taste in today's China is: first, local classicism-a scenery way by adopting strange stones and big trees-a way of robbing the nature. The other one is the western classics style-known for its exquisite techniques-well trimmed trees, dynamic fountain, marble floored yards-demanding great maintenance cost and resources. All these are controversial to the sustainable development philosophy.

The taste in our time needs to base on the ecology, while covering the taste of all things. Beauty is the balance and harmony. The time calls for the countryside, weed and even corroded old things. They are sustainable.

WHO IS A SCENERY DESIGNER

A scenery designer needs to stick to the land, just as a scholar puts it, a person handles the relationship of land and people. A scenery designer is responsible for rebuilding a beautiful homeland.

A scenery designer needs to do his job carefully. Otherwise one day our limited land will collapse and that is unimaginable for the whole world.

A scenery designer should not neglect the response. He should be predictable, standing side by side with human being, rather than one designing static things.

This writer would like here to quote a remark from a US-based scenery designer: "I don't think we can always make charming sceneries. I think we should make in-depth and significant sceneries. A building is just like one night stand, which is very hot and sexy. But it is just one night. A perfect scenery is like marriage, an experience which is in-depth, significant but always on the way to the mountain peak."

Landscape design of E' yangshan Park, Changsha

长沙鹅羊山公园景观设计

福地健身区全景

化挖掘、运动场馆建设和植物景观营造，使其成为运动、健身、生态、文化、旅游、休闲于一体的运动型生态文化公园。

总体布局与功能分区

运动公园的总体布局因地制宜，采取规则与自然相结合的布局形式，组织多层次空间。规划时对原有水体加以开挖、改造形成自然的湖泊，组成动态、静态、开阔、曲折等一系列景观，与周围的山、水、石、树木等自然物统一、和谐、融为一体，并收到〝虽有人作，宛自天开〞的效果。

根据公园规划的地理位置以及项目设置内容，从最大限度利用公园游憩机会的角度进行功能分区、景观分区和景点设置。从游憩机会序列和景观特色上将公园分为6个功能区，并设置有生态保护区、福地运动区、大地艺术及极限运动区、福地健身区、滨江活动区、福地森林游憩区、福地养生区、生态保护区。

E'yangshan is one of the 72 lucky places of the Taoism. E'yangshan is located at the east side of the Xiangjiang River and faces the Gushan across the river. There is an island in the river named Moon. The three sites form the first scene in the north end of Changsha.

Villas around E'yangshan demand highly on the environment and sources. The E'yangshan garden is designed to protect the resources while allowing people to live there.

Positioning of the theme park

Rate the traditional teaches of the Taoism and the modern life, the theme should be harmonious between men and the nature. The conception of the positioning is the original land.

Interpretation of lucky place and lucky place movement

The designers employed E'yang 8 Scenes to express the idea of lucky place, and the 22 scenes imply the order of E'yangshan in the entire 72 lucky places. This is E'yangshan park's lucky place concept, which makes the park a dynamic place of culture.

The lucky place movement means that people doing physical exercises in this place will be satisfied physically and emotionally. The E'yangshan park highlights the traditional culture characters of people heritage of the ancient city, expressing the demands of city people on parks.

Entire arrangements and functions

The arrangements consider both nature and handworks, making the space more diversified. The previous rivers were trimmed, forming some lakes supported by series scenes. These projects integrate with surrounding hills, rivers, stones and trees.

The project has made full use of the sources in doing the function and scene divisions and scenery arrangements.

1. 福地养身园区
2. 福地森林游憩区
3. 滑草场
4. 水上剧场

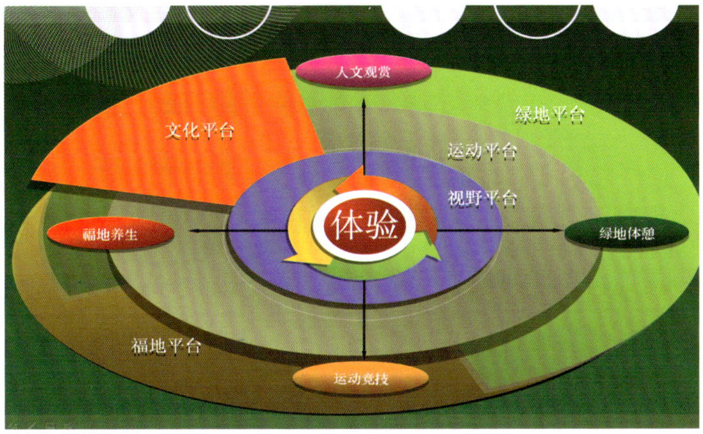

什么是"福地运动"

"运动"解读

工程档案

工程名称
长沙鹅羊山公园景观设计

工程地点
湖南省长沙市

设计单位
湖南省建筑设计院景观设计研究所

设计师
王小保、彭丹、唐贤巩、张剑飞、许海南、范米娜、曾丽丽

项目景观总占地
70.6公顷

单位平米造价
150元

设计完成时间
2007年8月

竣工时间
在建

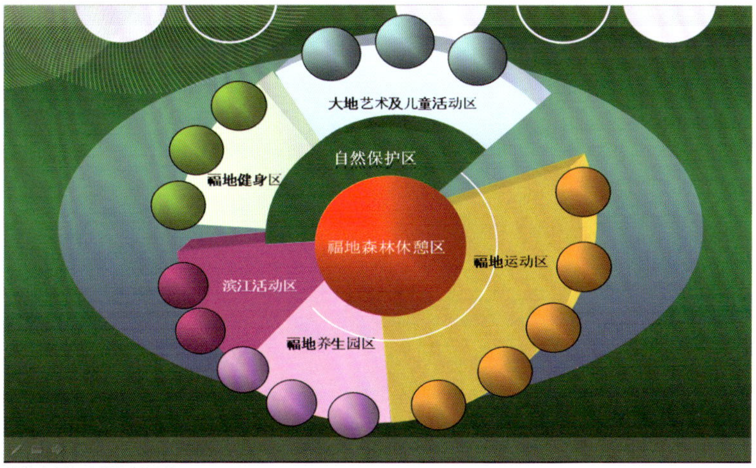

总体布局与功能分区示意

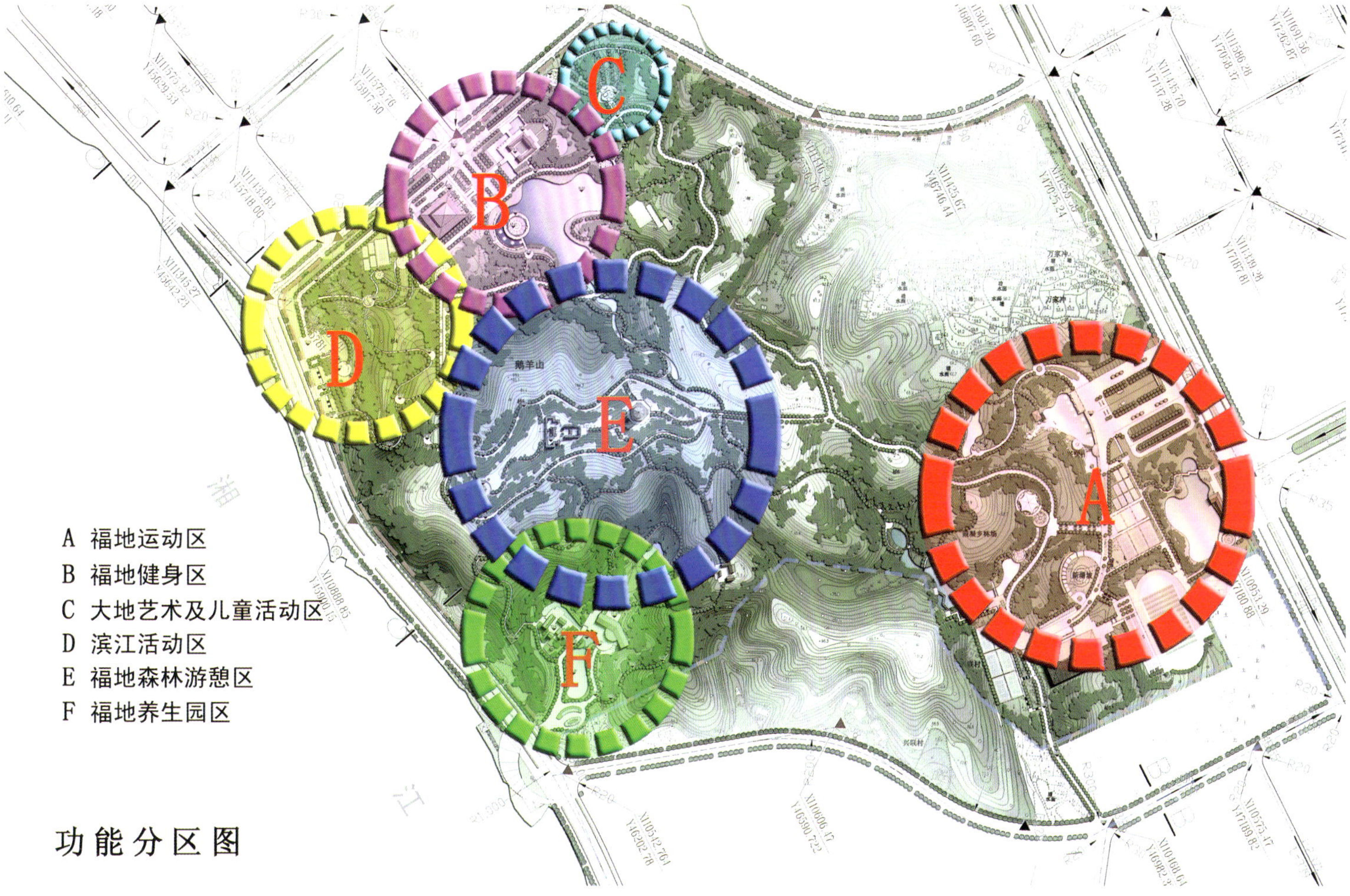

A 福地运动区
B 福地健身区
C 大地艺术及儿童活动区
D 滨江活动区
E 福地森林游憩区
F 福地养生园区

功能分区图

Landscape design plan of Yueliangwan Park, Quzhou city

衢州月亮湾公园景观设计方案

设计概况与目标

月亮湾公园位于衢州市西区，属于衢州新开发的区域。地块周边以居住区为主，地块东、西侧为待开发的居住用地，北侧为已建学校和住宅区，南侧为现有开关厂。周边地势平坦，北侧可借远山。月亮湾公园作为一个社区公园应该充分考虑到区域内居民的活动、休闲要求，强调空间感知和景观艺术性，使之成为社区居民的精神文化场所。

现状分析

地形地貌：基地地形由丘陵和平地组成，约15m的高差，整个地势由西南向东北方向渐渐降低。

土壤：基地的土壤为红壤土类的红砂土及少量的黄泥土。土壤粉沙性强，黏结力弱，土层浅薄，有裸岩。

植物资源：植被贫乏，以马尾松林为主，品种单一。东侧有坡耕地，次生的杂灌林与地被植物较繁茂，层次性和垂直密度低，基本无大树、古树。

场地特性：在山坡的东北侧有两处裸露的岩石，可加以保护性的开发利用。

水文：基地西侧地势较高，东侧农田存在低洼积水和灌溉渠。衢江有一支流石梁溪流经西区，但与基址较远。

山水景观构建与功能融入

山水景观构建：根据基地丘陵与耕地的现状，在两种地貌的结合部位，我们嵌入一条水系，在场址中部形成"月亮湾"。月亮湾是公园的山水景观核心，与山地裸岩共同把衢州当地的山水意向表达的淋漓尽致。

功能融入：基址的西部为丘陵，为山林区，同时也作为水系的背景，东侧为公共开放空间。西侧山林区融入静态功能，而动态活动则安排在东南面的公共开放空间。

基地东侧界面处理：基地东侧是待建居住用地与公园交界。建议在居住用地内划出20m的绿化带做控制带以形成自然过渡。

功能分区与节点设计

静态休憩区

位于基地的西、南部，带状的丘陵密林带作为公园的绿脉，为人们晨练、散步、爬山、品茶、打牌等提供活动场所。此地环境幽静，植被茂密，创造了亲近自然的空间体验。

西入口晨练广场：该场地位于紫薇北路东侧，呈南北带状分布，因现状土方开挖严重，设计时对地形做了处理，形成台地式的带状晨练广场，为居民提供安静的林下晨练空间。晨练广场与城市道路间以香樟、石楠、夹竹桃形成常绿隔离林带，晨练场地中种植冠大荫浓、树姿雄伟的珊瑚朴，形成林荫晨练空间。自然山林区：位于公园的西侧，利用原地形地貌，改造现有的马尾松林，增加常绿阔叶和秋色叶树种。形成以体现秋景为主的风景林带。山上设有游步道和休息亭廊，秋天满山的红叶成为一道亮丽的风景。岩石花园：设计时对山坡东侧颇具观赏价值的裸岩进行整体性的保护，设置一岩石园区，保留原有的富有野趣的岩生植物景观。利用山上

土生土长的植物资源，如金樱子、野蔷薇、算盘子、景天科、禾草类及苔藓类等，增加薜荔、络石、金银花等山区石壁常见的攀援植物，形成具有浓郁的当地特色的岩生植物景观。

动态活动区

位于基地东南侧，沿月亮湾展开，其核心为宁静的水面和临水的活动大草坪。可进行小型球类、溜冰、晨练、放风筝等活动。

休闲茶室管理区：在南侧山角处建造一茶室，结合最南侧的一片裸岩，营造一处由山林建筑围合的岩石花园茶座区。石滩垂钓区：在南侧山角下设置一片卵石滩，石滩缓坡入水，片植禾草类植物，形成富有野趣的水生植物景观，方便人们在此垂钓。草坪活动区：临水的大草坪，给人们提供了休息、野餐、露营的空间。沿居住区一带营造密林隔离带，也使草坪活动区更具有围合感。

总平面图

草场活动区详图

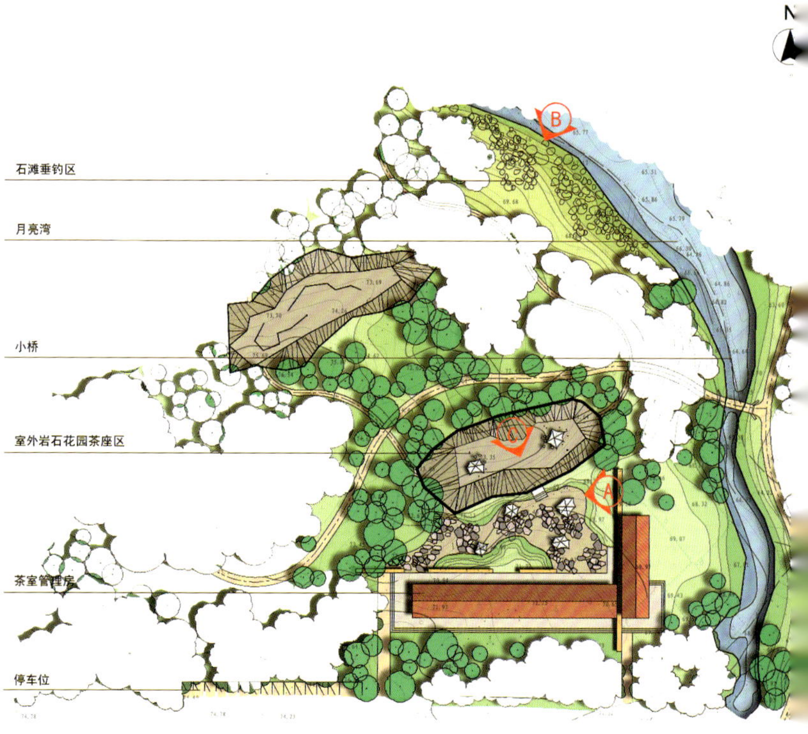

休闲茶室管理区平面图

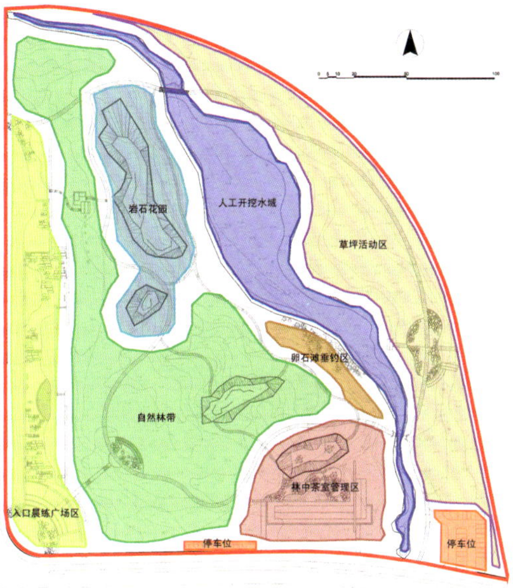

工程档案

工程名称
衢州月亮湾公园景观设计方案

工程地点
浙江省衢州市

设计单位
浙江大学建筑设计研究院风景园林分院

设计师
林建海、李辉、查丽君、李娴

项目景观总占地
6.89公顷

单位平米造价
320元

设计完成时间
2006年9月

竣工时间
在建

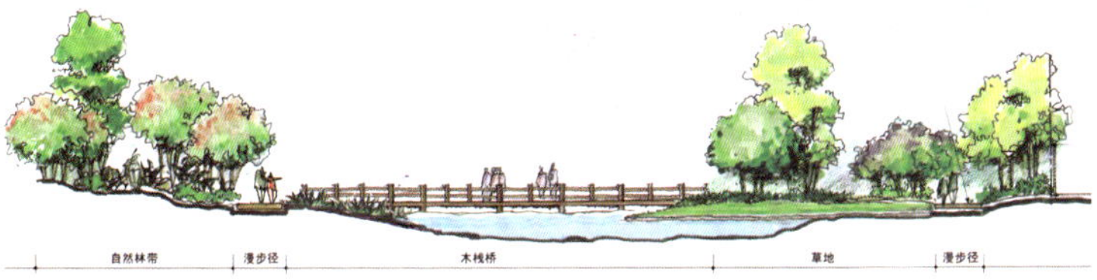

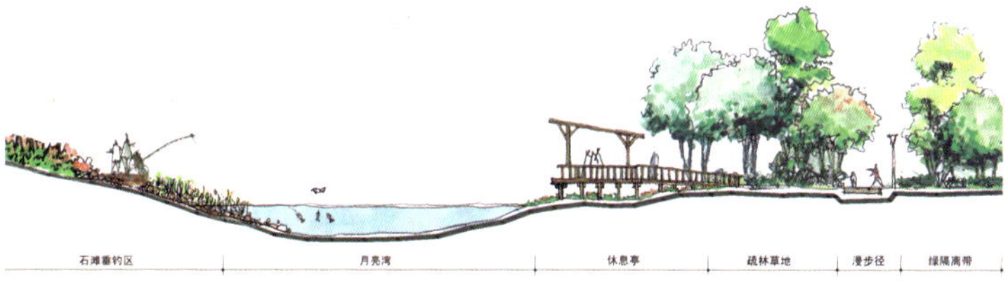

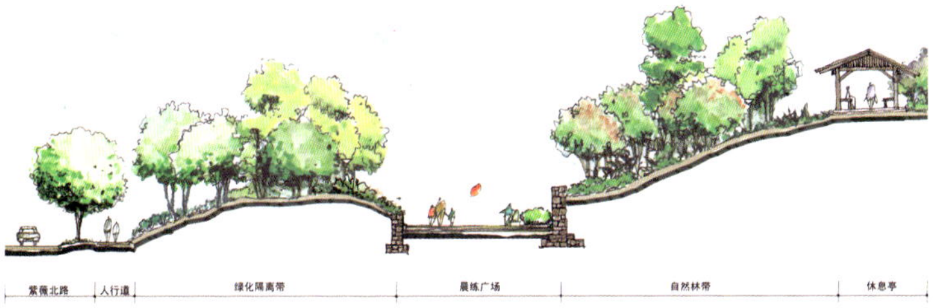

草坪活动区

DESIGN BRIEF AND OBJECTIVE

The Yueliangwan Park is a newly developed area, located in the western part of Quzhou city. The land is flat and there is a mountain at its far north side. Its surrounding lands are mainly for residential use.

CURRENT SITUATION

Landform, there are hills of 15 meters in altitude. The entire land goes down from southwest to northeast.

Soil, mainly red sand and less yellow clay. Bare rocks can be found on the flimsiness land.

Plant, mainly Chinese red pine. There are shrubs at east side land and no big or ancient trees.

Land character, there are two bare rocks at the northeast side of the hill.

River, there are some hydrocele and feed canal at the east end.

CONSTRUCTION OF HILL AND RIVER SCENERIES AND FUNCTION IMPLIES

Construction of hill and river, a river is introduced at the connection part of the land, forming a moonlike lake. The lake is the centre of the park.

Function implies, hills at the west end could be the background of the river and the east end is an open part. Static functions are implied in the western end hills while dynamic activities put at the open part of southeast end.

East end trim of land, and land is reserved for residential use and therefore a 20 meter wide greening belt is recommended as transition.

FUNCTION DIVISION AND NICHE DESIGN

Static area

Located in the west and south ends of the land. Quiet environment and opulent plants are good for leisure activities.

Sports plaza of west entrance, The belt-like plaza provides an area under the shadow of trees. A greening belt is between the plaza and city roads. Forest area Located at the west end of park, features winter sceneries. Walking paths and booths available. Rock park To make use of the existing resources, and plant more local plants.

Dynamic area

Located at the southeast end of the land, alongside Yueliangwan. Water and riverside grass are available for leisure people.

Tea house Located at the south end of hills, surrounded by a rock forest. Forest rock fishing area Located at the south end of the hill.Grass land area Riverside grass land is reserved for clinic use.

1. 茶室管理房模型
2. 茶室管理房
3. 晨练广场

西立面

东立面

南立面

北立面

3

Landscape design plan of Caojialong Reservoir, Quzhou city

衢州曹家垄水库景观设计方案

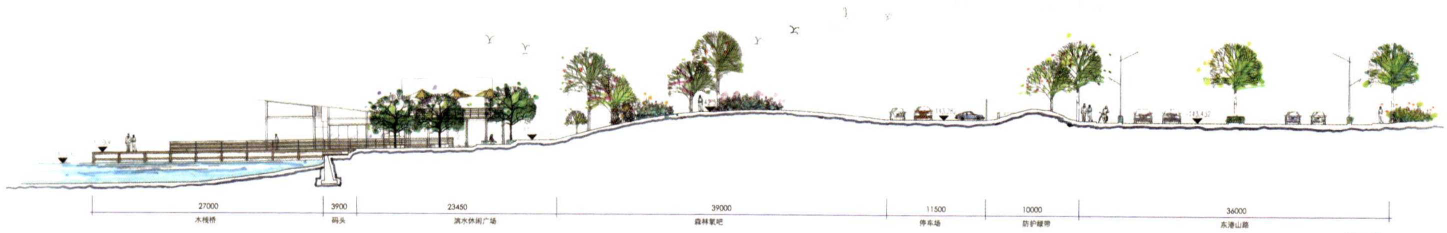

A-A 剖面图

本项目位于衢州市东港工业园核心服务区内，是展现园区整体风貌的重要窗口。水库公园以北的大面积土地已经平整，用于工业项目开发。水库公园的现状基本处于原始状态，以连绵的龙岗丘陵为主，相对高度变化不大，有20米左右。水库周边的山体地形较丰富，山体多岩石，林地多杂生灌木，少乔木层。山坳平地处以大量的桔园为主。水域面积约52000平方米，水质清澈，与地形结合较好。我们将此项目的设计定位为：高度开放、环境优美的环湖公园。通过营造核心区中心绿色、大气、雅致的环境氛围，建立一个全新的地标式景观，提升城市的个性与品位，为群众休闲提供一个好去处。

主入口广场

主入口广场是重要的休闲活动场所。相邻规划道路与水库水面的标高存在近5米的高差，利用这一特点我们在入口设计了一个特色水景：叠水坡，并与商业街入口的水池结合，既是水库补水的出水口，又是一道亮丽的风景。入口的广场采用台地式布置，解决了高差，也丰富了空间，主入口道路沿着山脚布置。岸线的处理上采用折线与自然相结合，延长了亲水岸线的长度，形成一串丰富多样的亲水空间：自然的草坡入水、临水的木平台、与水景结合的阶梯式亲水河埠头、穿梭于水面的木栈道。一条堤坝横穿水面，将广场与对岸的景点连接。两侧成片种植挺立的水生植物，将堤坝隐于一片绿色之中，在接近山体的地方架起一座精致的小石拱桥。人游走其中，可体验到别样的空间感受。

河畔餐饮区、梯田休闲区

梯田休闲景观区利用高差形成梯田式的特色植物景观，人工铺设的休息场地和观景长廊穿插其中，上层种植景观树。

1~5. 临水餐厅效果图

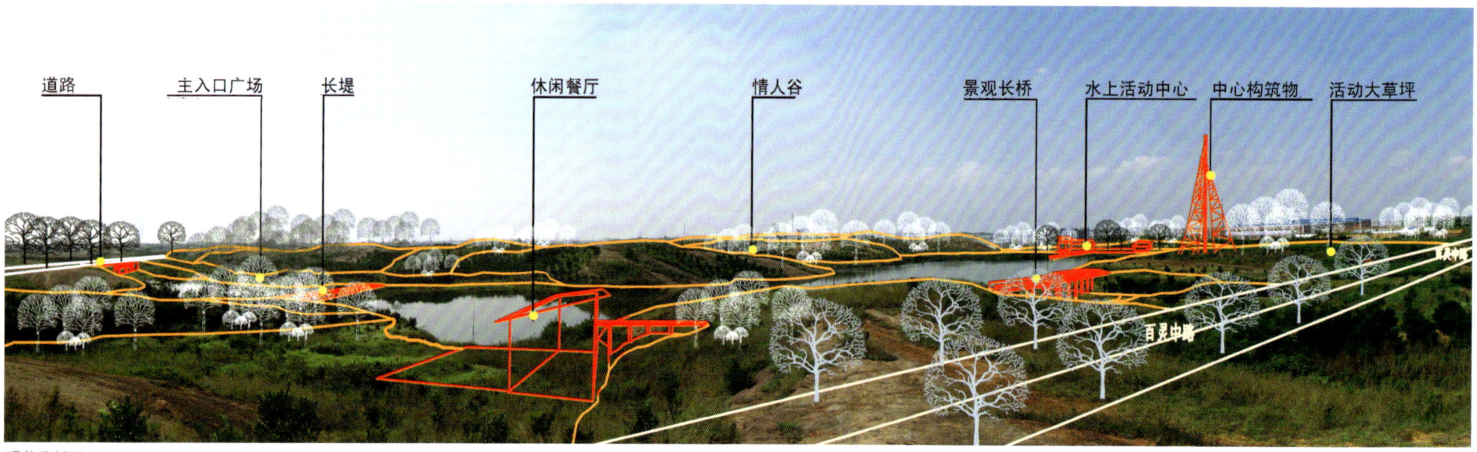

现状分析图

河畔餐饮区位于两个山谷之间。休闲餐厅的建筑设计采用了两层的单坡顶建筑，一侧架空于水面，背面紧靠裸露的岩石山体，在二层架了一条通道，直通山顶的岩石休息平台。建筑主要以钢、石材、玻璃为主要材料，简约、空灵又与岩石山体、树林、水面和谐相融。这里是商务宴请、朋友聚会的好去处，环境优美、私密宁静。

大草坪活动区

此区块是整个公园主要的休闲活动场所，地块两面临水，东侧紧靠百灵路。完整的草坡由东往西延伸至中央水景区，与道路和垂钓区交接的部位密植观赏性的林带，形成一定的围合感。大草坪的南侧保留了一长条临水的裸岩，裸岩的肌理和色彩较好，此处架一条高出水面3米多的景观长桥，直通向对岸的岩石山体，成为一处吸引视线的景观焦点。草坪的中心处设计了一个高35米的景观构筑物，形成公园的

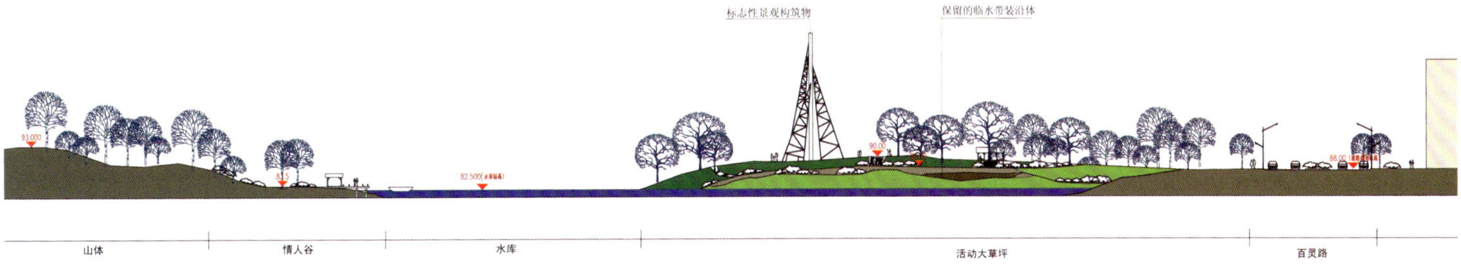

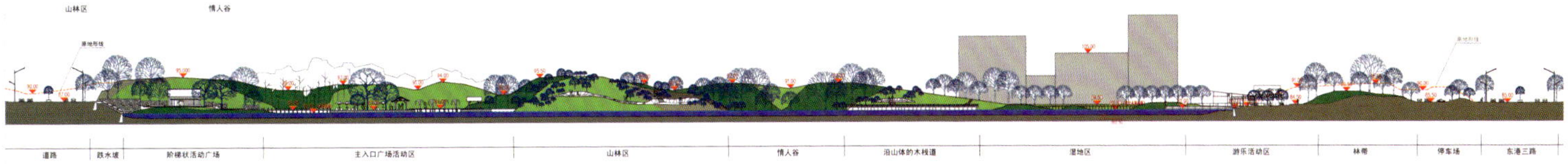

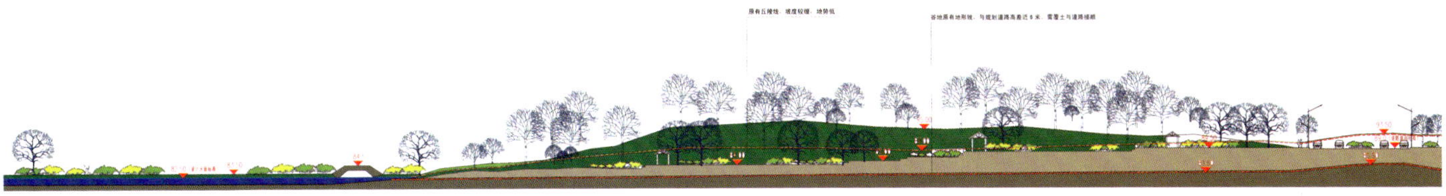

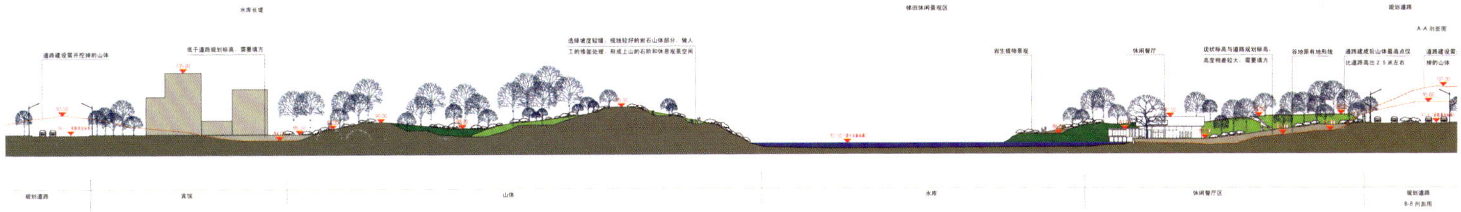

自然驳岸

湿地小岛

栈道式驳岸

裸岩直接入水

直立式硬质驳岸

中心控制点，也是核心服务区主轴线的视线焦点。

游船码头区、垂钓俱乐部

游船码头区块主要功能空间均分布于临近湖岸处，形成滨水休闲空间及防护性疏林草地。改造原有的鱼塘形成垂钓区，结合树林草地以及林中的建筑群，打造清馨雅致的休闲景观。

湿地区

在原有堤坝的周围规划了一片岛状循环湿地，雨水在流入河道之前先经过湿地的净化，可改善水质，同时富有特色的湿地景观也为城市提供了一道美丽的植物景观。低矮的草本水生植物，如芦苇、茭白、香蒲等形成富有野趣的水中小岛。沿岸线则选用菖蒲、再力花、海寿花、矮生美人蕉、水葱、花叶芦竹、睡莲等形成精致的水景，供游人近距离观赏。

情人谷

情人谷位于一处临水的山谷，得天独厚的地理位置，为营造浪漫宁静的景观奠定了基础。这里以纯粹的植物景观为主，一条简单的原路穿梭于树林、草地、水湾，沿着原路布置有座凳，长廊，游船码头。

岩石景观

在大片裸露的岩石山体表面有几个观景视点较好，且坡度较缓，适合游人停留的地段。通过人工的修凿、挖掘，塑造出一些富有个性的岩石景观空间，可放置一些石臼、石槽，种植耐旱且生命力顽强的岩生植物。这些植物或丛生翠绿或鲜艳美丽或外形奇特，形成独特的岩生植物景观。游人可在此停留休憩、观景，也可把这作为烧烤的绝好场所。

1

2

模型示意图

可烧烤，游憩

种植岩生植物

挖掘下沉的岩石通道

观景的平台

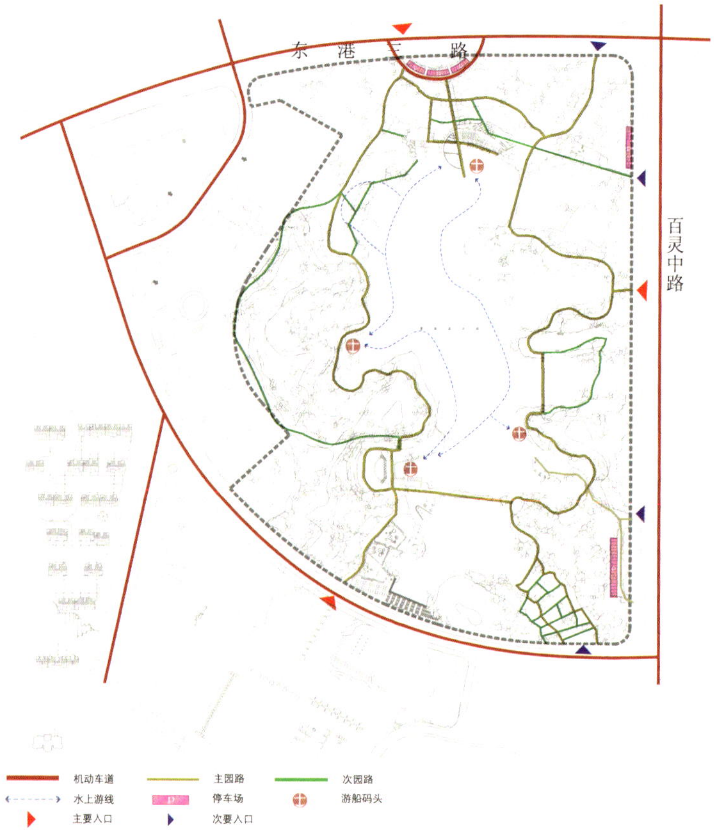

1. 长堤效果图书馆
2. 临水广场
3. 湿地栈道效果图

3

The project, a window for displaying the attractive image of the hosting city, is located in the core service area of Donggang industrial zone, Quzhou city. In the northern of the reservoir park is a large piece of flat land for industrial projects. The primarily reserved park is full of hills, with an altitude height of around 20 meters. Around the reservoir are rock mountains, a lot of shrubs and fewer tree stratum. There are a lot of orange parks within the laps too. The high-quality water area is 52,000 square meters wide combines well with the landform. The project defined as a highly opened and beautiful lake garden. The green and elegant core area will become a new landmark to upgrade the city image and provide a place for the leisure activities.

THE MAIN ENTRANCE SQUARE

The square is designed as a place for leisure activities. Ways around the square is almost 5 meters over the reservoir level. A fall water scenery is designed here to join the water pool at the entrance of the commercial street to supply water to the reservoir.

There is a tableland at the entrance square to solve the altitude problem and enrich the space. The roads are designed alongside the mountain foot, supported by a combination of broken and natural lines. A series of waterside sceneries help enlarge the waterside lines, including water emerged grass, wooden platform, stepped water pier and the wooden plank way. A dam goes cross the water, connecting and square and the sceneries on the other side. The water plants make both sides of the dam green. There is a stone arch bridge close to the hills and people walking inside the park will have a different feeling.

RIVERSIDE CATERING AND TERRACE AREAS

The terrace area accommodates plants, with artificial rest areas and scenery corridors crossing them.

The catering area is located between two valleys. The two-storey single slope dining hall stands on water in one side and backs a nude rocky mountain. There is a corridor on the second floor, leading to the rest platform on the top hill. The materials include steel, stone and glass which are made compatible with the environment. This is a good place for business banquet and friend reunion.

GRASS LAND AREA

The area, a leading place for the leisure activities, is surrounded by water in its both sides and a road in its east side. A complete grass slope extends to the central water scenery area. The connecting area of the road and fishing area will accommodate forests to make people here feel safety. At the southern end of the grassland a piece of nude rock, with good grain and colors, is reserved. A scenery bridge 3 meters over the water will be built to the rocky mountain on the other side. A 35 meters high building is designed in the centre of the grass land to become sight spot.

BOAT PIER AND FISHING CLUB AREAS

The boat pier areas are located in the waterside areas, creating leisure space and protective grassland. The fishing area is transformed from the previous pond.

WETLAND AREA

An island shape cycling wetland is designed around the existing dam. Waterfalls will be purified by the wetland before running into the rivers. The wetland scenery is so precious for the city. Reed, water bamboo and cattail help form a water island. Other plants alongside the water include flagleaf, canna and candock.

LOVER'S VOLLEY

The volley is located in a waterside volley. A simple road goes through forests, grassland, water falls, supported by chairs, corridors and piers.

ROCK SCENERIES

There are several places on the rock mountains, with better eye point and grade for people stay. Some artificial sceneries can be made, joining the stone mortar, slot and some rock plants. This is a best place for the rest, sightseeing and barbecue activities.

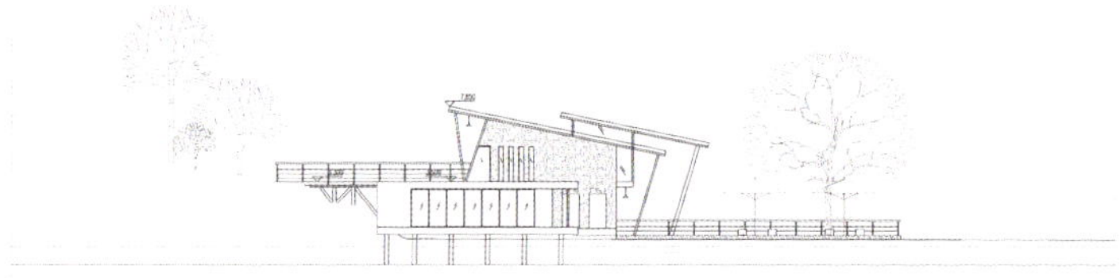

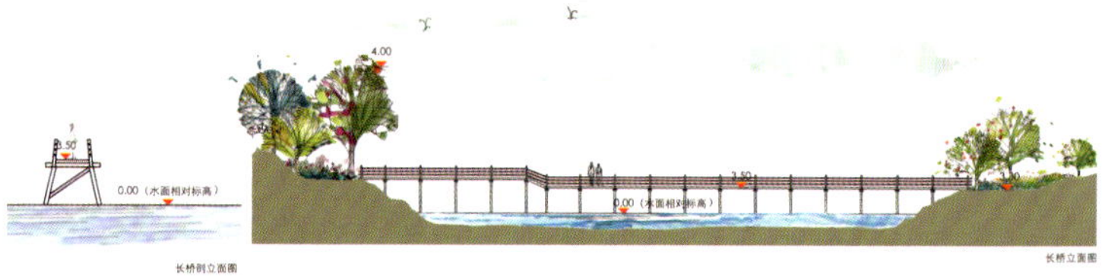

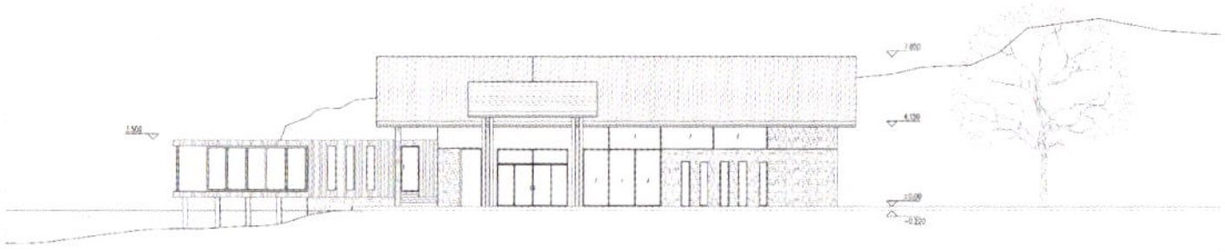

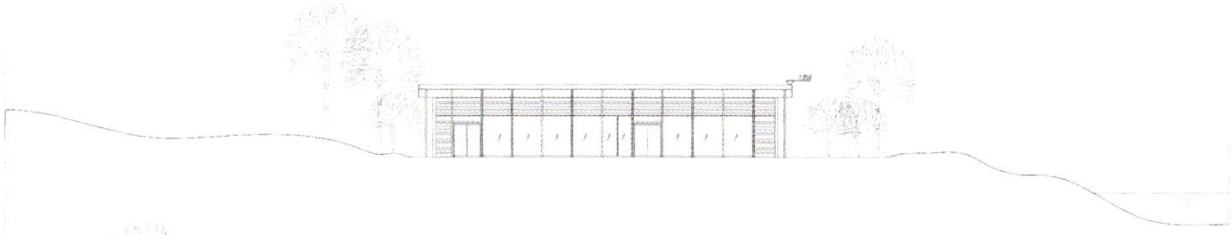

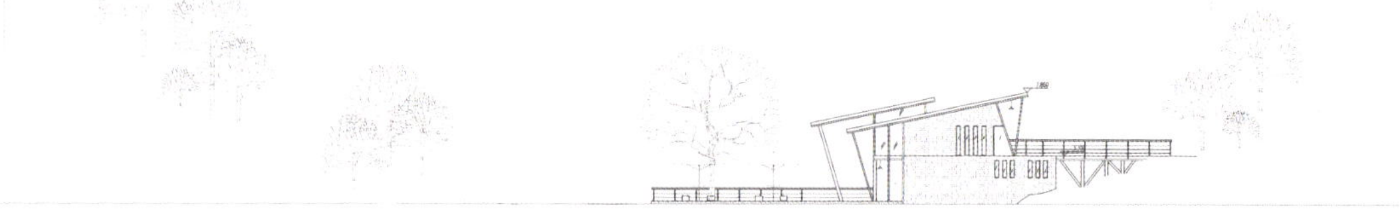

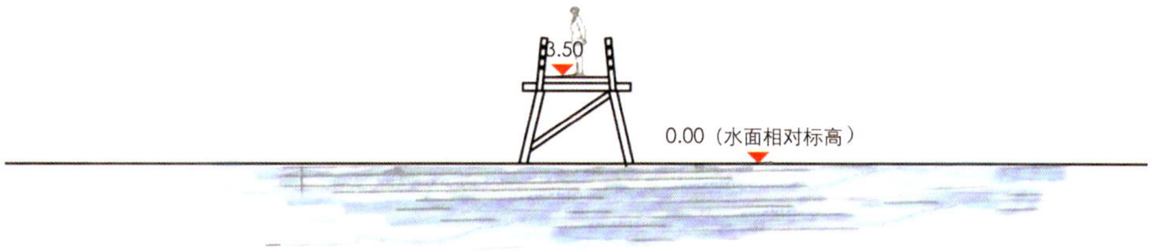

长桥立面图

工程档案

工程名称
衢州曹家垄水库景观设计方案

工程地点
浙江省衢州市

设计单位
浙江大学建筑设计研究院风景园林分院

设计师
查丽君、李辉、汤珏、李文江、刘军峰、李瑛、张燕

项目景观总占地
17.9公顷

设计完成时间
2008年2月

1. 大草场效果图一
2. 大草场效果图二

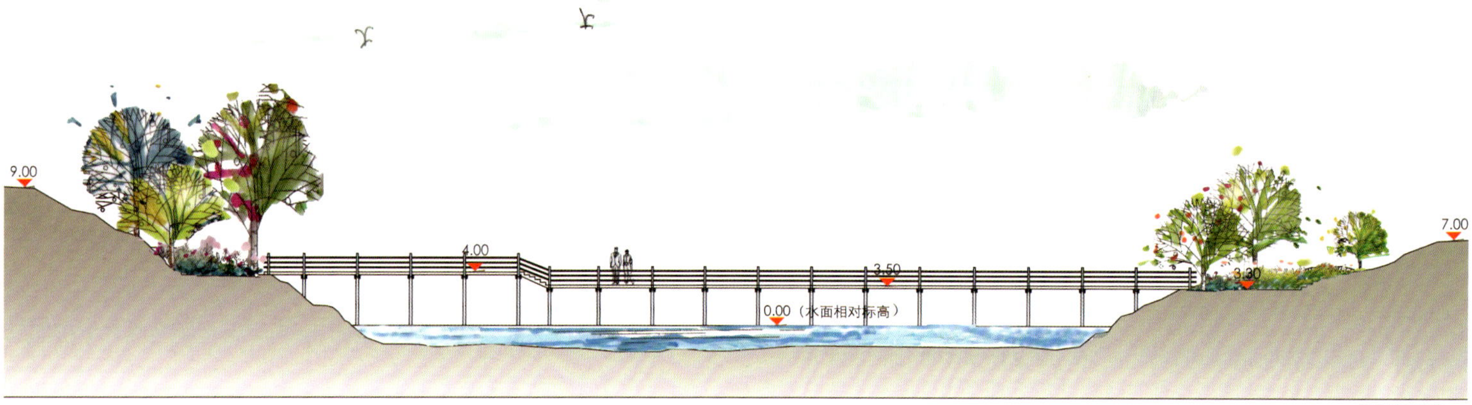

长桥剖立面图

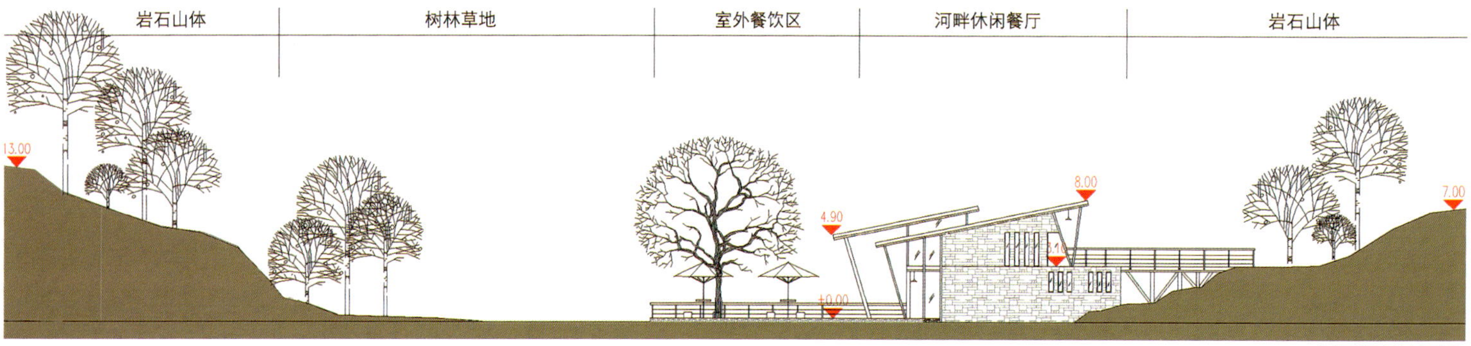

休闲餐厅南立面图

1

2

Landscape design of Binjiang Park, Nantong city

南通滨江公园景观设计

1

2

3

1.3. 滨江商业街
2. 书画馆

现状概况

本项目位于南通狼山风景名胜区西北部，北邻狼山港区，南抵龙爪岩景区，属北亚热带气候，因地处长江三角洲北岸，濒临黄海，故海洋性气候特征明显。本区地属长江中下游冲积平原，地形平坦，地块东侧为新围垦区，地势较低，除小部分堆土区标高为5.4m～6.8m外，大部分为江滩地，标高为-1.3m～1.2m。地块西侧为大片湿地，河网水塘面积较大，大部分地面标高为2m左右。

布局规划

根据周边环境、地形地貌以及反映南通城市特色等方面对滨江公园进行规划，将公园分成五个景区：港口风情区、江岸风光区、碧水金沙区、休闲天地区、湿地生态区。

港口风情区：公园北邻狼山港区，区内巨大的吊车成为公园西北天际线上最醒目的"风景"，用树木或其他方法均无法遮挡。因此本方案设计因势利导，将港区的远景和公园景观巧妙结合，视远处高大吊车为借景，组合公园内部的"激光塔""港桥""金

1

锚”“老船长”等主题景点，将南通这个重要的长江港口城市的特征表现得淋漓尽致。高高的“激光塔”成为滨江公园的标志。

江岸风光区：滨江公园沿长江景线长850米，江岸风光区由“亲水平台”“波浪平台”“克莱特纪念像”“江上世界”“凌波台”及沿江各小品组成。“亲水平台”从江堤顺台阶伸入江面近30米，即使在枯水季节也能保证台阶自然伸入江水中。“波浪平台”是从江水波浪中获得灵感而设计的亲水平台，从江中看“波浪平台”大堤仿佛是层层波浪向前推进，它既起到了亲水效果，又改善了江堤单调的形象。

碧水金沙区：滨江公园中心位置上设计一个音乐喷泉泳池，该泳池由旱喷泉、山林瀑布、大小三个嬉水池及通往泳池的斜拉桥等组成，泳池边有从南海运来的金色珊瑚沙铺成的沙滩，夏暑季节人们在此沐浴江风、水中嬉戏，在不能游泳的季节，泳池内的喷泉随音乐翩翩起舞，是人们休闲观光的好场所。

休闲天地区：休闲天地区由一组现代休闲式景观建筑群组成，该建筑群结合地形高差顺坡而建，并形成“夹水”“夹道”格局，建筑群对内可观湿地水景，对外可观金沙滩和远处江景，这里既是滨江公园的服务中心，也将成为南通市休闲生活的一个时尚中心。

湿地生态区：区域内现有河网及若干水

1. 金锚广场远眺
2. 总平面图
3. 波浪平台景区
4.5. 商业街
6. 纪念像广场
7. 露天小剧场

工程档案

工程名称
南通滨江公园景观设计

工程地点
江苏省南通市

设计单位
中国美术学院风景建筑设计研究院
杭州杰顶环境艺术设计有限公司

设计师
钱江帆、秦瑛、刘超、吴维凌、李剑峰、郑炜

项目景观总占地
26.7公顷

单位平米造价
400元

设计完成时间
2006年1月

竣工时间
2006年12月

3

4

5

6

7

塘，设计中尽可能地保护这些宝贵的湿地资源，并且在原来的基础上适当整治，形成特有的“原生态水景”。方案中将排灌用的水利工程设施经过“包装”和组合，形成一处新的有历史遗存感的景点。

本方案还在地块南部靠近龙爪岩景区处设计一座露天音乐广场；在和龙爪岩景区结合部设计一座书画馆，这座传统形式书画馆的建设既充实龙爪岩景区的内涵又延续了龙爪岩景区已经形成的建筑风格。

绿化设计

根据南通北亚热带气候和当地植被生长的特点，对公园绿化设计提出如下原则：绿化要符合时代特征，体现自然、大气的风格，注重生态平衡的原则，形成稳定的植物群落；绿化要符合地域特征，以本土特色树种为主，形成四季常绿景观，同时强调季节变化的特色；种植设计要体现总体布局的功能要求，同时重视生态效益。要统一全园区的植物风格，确定绿化的基调树种，每个区域根据其形成景观的要求确定合适的植物种类，既要有各区域的特色，又和整个公园的氛围相协调，达到林成荫、树成群、草成坪、花成片。

道路设计

滨江公园主道路系统近似“田”字形，南北向三条景观主干道分别以三条堤坝为基础进行适当美化、调整，并结合多个小广场、休闲平台等，将线性路网与若干个面结合。最东面的竖向干道为双边密林式林荫道；中间干道为单边密林为主的休闲道；最西边沿江道为宽敞开阔的滨江景观道。东西向三条景观主干道分别是：一条位于区块北部的以“金锚”雕塑为终点的景观轴线道；一条位于区块中部，穿越休闲天地通向亲水平台的道路；另

1. 激光塔
2.6. 儿童沙滩
3. 波浪平台
4.5. 湿地水塘

一条位于给水博物馆北通向江边。另外，在主干道骨干的基础上布置联系主景点与次景点、次景点与次景点之间的次干道和庭院道，这些道路的设置能够使游园的人们通过迂回曲折的道路游赏各个景点。

灯光照明设计

根据公园内不同景点和功能的特性采用不同亮度和形式的照明，主干道和各主要广场一般为金属卤盐反射灯结合草坪灯，灯高3～6米，次干道采用园林灯或草坪灯，园林灯高一般在2.5～3米。主景点、大树、建筑物有大量投射灯照明，激光塔内部有灯光透出，上部有激光射线，主干道、广场地面、亲水平台有地埋灯，喷泉下面有彩灯，休闲天地区树林有成组投射灯，建筑物有轮廓灯、霓虹灯。

CURRENT SITUATION

The project is located in the northwest part of the Nantong Langshan Mountain landscape area, neighboring the Langshan port area and the Longzhaoyan landscape area in its north and south ends. The land is part of alluvial flat of the Yangtze River and apart from partial land which is 5.4~6.8 m higher than the sea level, most part of the land is -1.3 m ~1.2 m of the sea level. A wet land is located in the west part of the project.

LAYOUT PLAN

The park is divided into 5 parts: port flirtation, riverside landscape, blue water and golden sand, leisure area, and wet land.

Port flirtation area. The area takes the high crane in the far end as background and other themes such as the Laser Tower, Port Bridge, Golden Anchor and Old Captain help remind people of the nature of Nantong city. The skyscraper Laser Tower is also a symbol of Binjiang Park and a transition of the port and park areas.

Riverside landscape area. The 850 meter long area consists of riverside

1

2

3

4

5

6

7

8

南通市滨江公园道路分析总平图

南通市滨江公园景观分析总平图

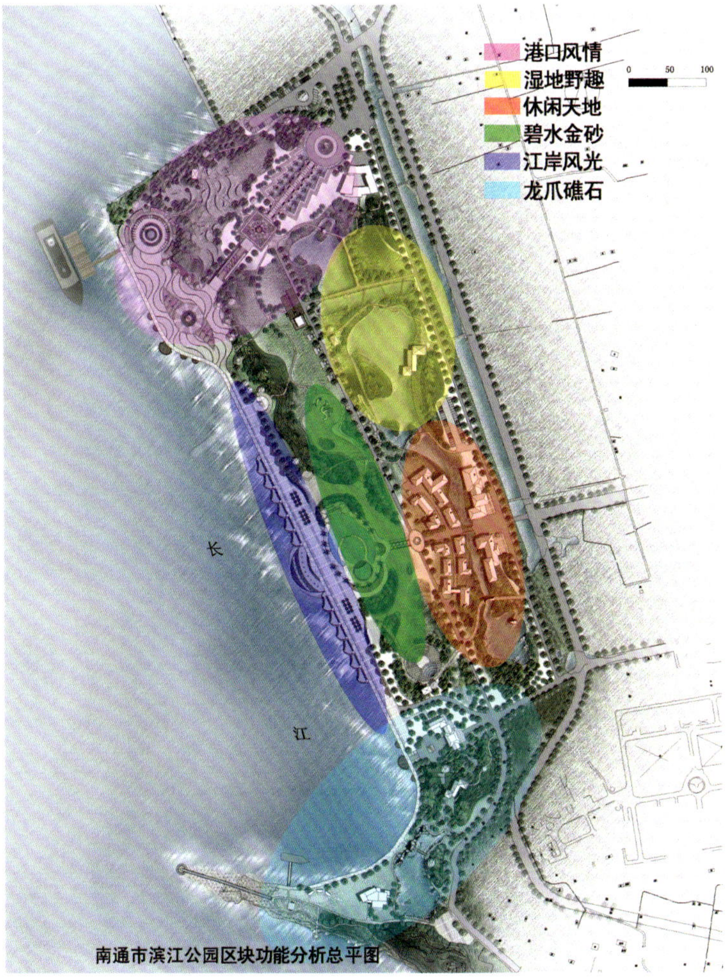

南通市滨江公园区块功能分析总平图

podiums and other water-related ornaments.

Blue Water and Golden Sand area. The center area of Binjiang Park houses a music fountain swimming pool which consists of several parts. The sand beach serves swimming and leisure people.

Leisure Area. The area consists of a group of modern leisure buildings, which survey both the wet land and the golden beach and other river sceneries. The area is the service centre of Binjiang Park and a fashion center of Nantong residents as well.

Wet Land area. The designers try to maintain the existing resources and these trimmed resources becomes a new historical site.

The entire design also includes an open music plaza and a book/ picture museum.

GREENING DESIGN

The principles are: the greening needs to show the time trend, featuring ecology balance and plants stability; to adopt more local tree types, showing green colors in the 4 seasons; each section has its own character.

ROAD DESIGN

The three south-north roads are based on the existing dykes, supported by multi leisure plazas and podiums. The upright road in the far east end is both-sides boulevard; the one in the middle is a single side leisure road and one in the west end is a riverside landscape road.

The 3 main east-west roads are: one axis path locates in the north part of the project, ending at the foot of the Golden Anchor; the second one is located in the middle of the project, crossing the path connecting the Leisure Area and riverside podiums; the last one is located in the north side of a museum, leading to the riverside. Also, some connecting and secondary sceneries are arranged alongside the main road, together with some secondary roads and courtyard roads. These paths make all sites accessible to people living there.

LIGHTING AND ILLUMINATING DESIGN

Various brightness and shapes are adopted. Over the main roads and plazas are reverberators and grass land lights which are 3~6 meters high. The secondary roads adopt parking lights or grassland lights which are 2.5~3 meters high. The main landscape sites, trees and buildings are trimmed by projection lamps. The laser tower is penetrated by lighting from inside and there is laser lighting over it. There are underground lights alongside the main roads, floor and waterside podiums. Other sites are also supported by various lights.

1~8. 商业街区

1

2

3

4

5

6

7

8

Landscape plan and design of Houtan Park, Shanghai

上海后滩公园景观规划设计

依据上海总体规划，上海绿地发展的主要内容为绿地、林地和湿地。因地制宜地保护、利用世博会会址内后滩处因黄浦江潮汐影响形成的一片天然湿地，是探索上海市中心城市发展与自然和谐共存的重要课题，为此按照“城市让生活更美好”(better city, better life）的世博主题，贯彻生态和谐、节能环保、资源整合的原则，处理好保护和利用的关系，自然湿地与人工湿地的衔接和融合成为我们此项目规划设计的依据。

设计特色

——一条蓝带串起的四种文明

蓝带：三带一区、三场九园、步道网络形成的总体结构。

三带一区

滨江芦荻带：场地北侧是由芦苇和荻草构成的生态滨河岸线，由布道穿越，自然而亲切。

内河净化湿地带：场地中部形成内河湿地，通过过滤渗透层与黄浦江相通，建立湿地生态净化系统，将河水净化，为世博园区提供景观用水。

梯地禾田带：利用千年一遇防洪堤与湿地的5m高差，形成高差错落的梯田，植五谷与经济作物，田埂为径，辟乔木林荫空间。

原生湿地保护区：完整保留场地内滨江原生滩涂湿地，建立湿地保护区，吸引鸟类栖息。

三场九园

三场：分别为西端“空中花园”广场（通过改造利用工业建筑形成），中部水门码头广场，东端漂浮的花园（与会展区相联系）。

九园：在场地中建立多个艺术“容器，”形成多重体验空间。

步道网络

以网络的形式构建步道系统，串联各个体验空间，形成便捷自如的人行交通系统。

四种文明：“滩”的回归、田园江水、工业遗存、后工业生态文化

“滩”的回归：保留滩涂，湿地，回归自然，追忆渔猎文明。

田园江水：借用农业元素，体验农业文明。

工业遗存：保留工业化时代遗迹，回顾工业文明。

后工业生态文化：构建生态系统，建立多重体验与开放空间，畅想后工业时代文明。

设计应对的问题与挑战

上海世博会园区后滩湿地公园的设计面临六大问题和挑战：

后滩地区保有上海市区、黄浦江边仅有的一块天然湿地，它历经了两千多年的历史变迁，目睹着黄浦江畔农耕经济的兴衰起落，见证近代民族工业的发展历程，注视着上海跃升为世界大都市的步伐，是一处具有丰厚历史文化底蕴的珍稀绿地。随着城市的发展，它必将成为人们追忆过去、关注现在，展望未来的体验场所。因此如何遵从生态设计原则、秉承场地文化脉络、满足多重体验需求成为设计面临的第一大问题。

湿地是地球上具有多种功能的生态系统，地处长江三角洲的上海，突出的特点就是沿江沿海大面积的滨海湿地、境内众多河流和湖泊等多样的湿地生境。但是上海的发展使得湿地生境受到不同程度的干扰。后滩的原生湿地就是这一现状的直接见证。如何保护、恢复、重建湿地生境，充分发挥湿地的自然生产力，实

景观结构分层分析图

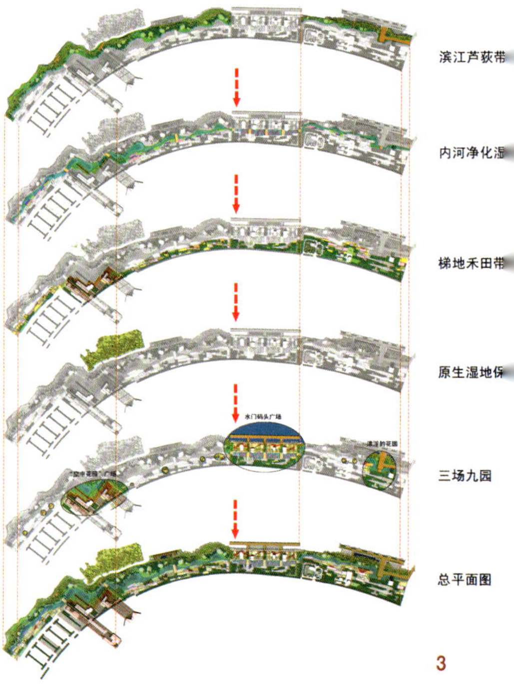

3

1. 总平面
2. 绸带
3. 景观结构分层分析图
4. 梯地禾田鸟瞰图
5. 水门码头
6. 芦荻台

4

5

6

1

2

3

4

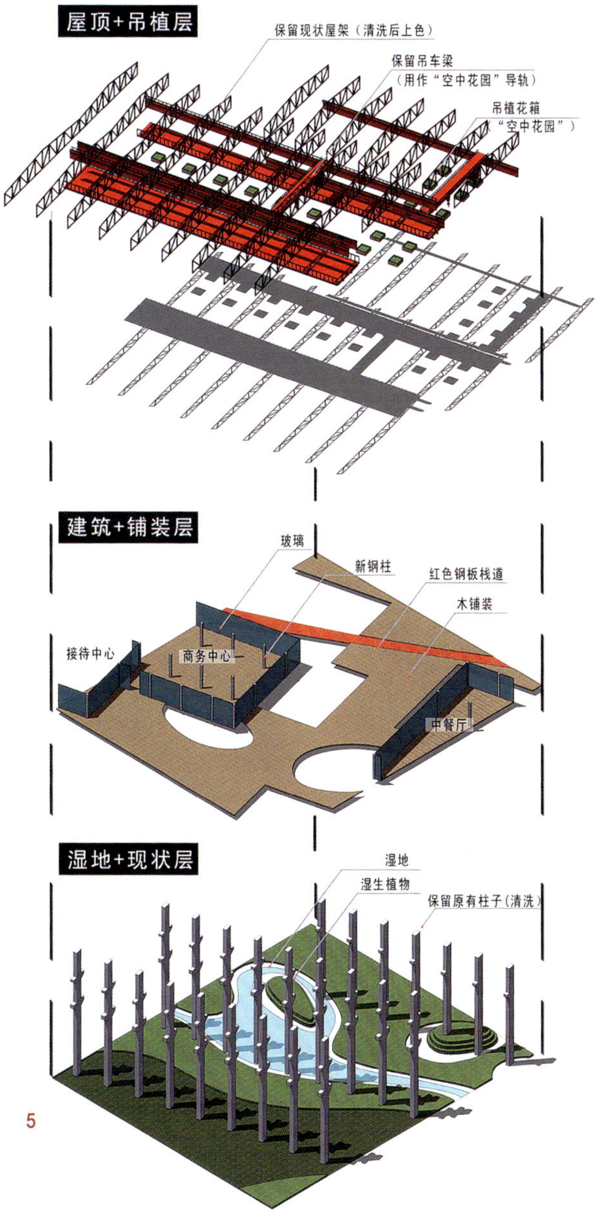

5

工程档案

工程名称
上海后滩公园规划设计

工程地点
上海市

设计单位
北京土人景观与建筑规划设计研究院

设计师
俞孔坚、凌世红、袁伟、龙翔、刘向军、孔祥斌、金园园、石颖、丁红霞、袁天远、蒋伟荣等

景观面积
18.2公顷

设计完成时间
2007年3月

竣工时间
在建

1. 水门码头
2.3. 梯地禾田
4. 漂浮的花园
5. 空中花园结构布局

现净化和调节水体等的生态功能，提高生物多样性水平，实现湿地景观的自然化，成为设计面临的第二大问题。

世博会期间人流量大且集中，设计日流量为40万人次，高峰日流量60万人次，极高峰日流量80万人次。而场地内的合理容量为2.66万人，极限容量为3.78万人，且场地南部临接世博园区西入口人流等候广场，场地中部的水门码头是世博园南区唯一的水上门户，场地东部则连接世博园区中心绿地，因此如何合理组织会时人流交通，协调好分流、等候与疏散的关系，成为设计面临的第三大问题。

依据世博园区总体规划的防汛设计要求，防汛墙的设计标高为千年一遇的防洪标准6.7米。而黄浦江的平均潮位为2.24米，平均高潮位3.29米，平均低潮位1.19米，防洪标准与黄浦江水位之间的高差多达5米左右，因此如何解决场地千年一遇防洪标准与湿地之间的高差问题成为设计面临的第四大问题。

世博会期间和会后场地的功能定位发生了变化。会时的功能定位偏向于安全疏散、游憩等候等功能的世博绿地，会后则突出湿地保护、湿地生态的审美启智和科普教育等功能。如何在设计时考虑相关弹性措施使其既能满足会时世博绿地的要求，又能方便、经济的转化为会后的城市湿地公园，成为设计面临的第五大问题。

湿地公园的设计以湿地的自然复兴和恢复湿地的领土特征为指导思想，应将湿地作为生物和能量交换的生态廊道，实现物质的自然循环。在作为探索城市发展与自然和谐共存课题的后滩湿地公园设计实践中，如何运用现代的理念和先进的技术营建出这样的湿地公园成为设计面临的第六大问题。

方案的六大设计对策

为综合解决以上六大问题，本方案提出六大设计对策：

第一：四种文明串写场地脉络

方案采用立体分层布局的方式，以后滩地区发展的时间脉络、空间背景和场地禀赋作为线索，将湿地公园分为湿地生态景观层、农耕文明景观层、工业文明遗存层和后工业文明体验层四个功能层次，由此叠加形成场地的总体功能布局。

第二：三带一区建构湿地基底

方案以“双滩谐生”为结构特征建立湿地体系，共有滨江芦荻带、内河净化湿地带、梯地禾田带和原生湿地保护区四个部分组成。其中滨江芦荻带和原生湿地保护区主要指与黄浦江直接相邻的外水滩地。外水滩地中的原生湿地保护区通过隔离保持其原生态的自然风貌，保护自然湿地免受人为干扰；滨江芦荻带则可大体分为砾石滩湿地和粗沙滩湿地两部分，它们共同完成过滤净化、防潮护坡等功能。

内河净化湿地带和梯地禾田带主要是人工湿地带，整体突出湿地作为自然气息地和水生系统净化的功能、湿地生态的审美启智和科普教育等功能。外水滩地和内河净化湿地带之间通过潮水涨落、无动力的自然渗滤进行联系，它们息息相关，一同营造着具有地域特征、能够可持续发展的后滩湿地生态系统。

第三：三场九园、步道网络编织交通体系

方案以一环九纵多路径的交通路网和三场九园的休憩场所共同形成场地的交通网络。

一环九纵多路径的交通路网，既确保了场地与外界的便捷联系，又保证了场地内部便利的可达性，很好的解决了世博期间人流的分流问题。三场九园的休憩场所，在形成不同文明景观体验空间的同时，提供了世博期间人流等候、疏散的场地。

第四：梯地禾田梳理场地高差

方案利用场地农耕文明景观层的梯地禾田来消解场地千年一遇防洪标准与内河净化湿地之间的高差。场地农耕文明景观层主要由梯地禾田带构成，它位于场地的南部，是场地与城市的过渡地带。梯地禾田带通过提炼

"田"这一特色景观，不仅消解了千年一遇防洪标准与内河净化湿地之间的高差，而且反映了场地近千年的农耕文明景观，有利于场地与城市的融合，丰富了场地与城市交接的景观界面。

第五：弹性设计满足双重需求

本着节约资源、会后尽量少改动的原则，在设计时充分考虑会后的利用，在功能建筑体和公共服务设施、铺装场地等相关方面进行会时与会后的弹性设计。

功能建筑体和公共服务设施等结合世博会时会后统一考虑，一方面采用可拆卸、可回收的材料，另一方面进行周密设计方便其功能转化定位；而铺装场地等因涉及面更广，可结合场地现状，选择相关可用于场地铺装的材料，如砖、石废弃物等，以及乡土材料如竹材等。砖、石废弃物等可先进行粉碎，然后在会时作为场地铺装材料，会后则可转变为绿地的透水垫层等。竹材在会时可作为临时铺装材料，会后则可拆除、粉碎作为有机肥料。

第六：生态理念引领技术航向

生态与人文理念贯穿于后滩湿地公园的全部设计过程，主要体现在再造湿地公园、场地工业遗存保护再利用以及其他生态可持续发展的相关措施与技术等。再造湿地公园的相关措施与技术主要有：湿地土壤、微生物过滤等新技术的运用。场地工业遗存保护再利用的相关措施与技术主要有：对工业厂房、构筑物等的保护意识和技术处理手段。其他生态可持续发展的相关措施与技术主要有：生态护岸与生态防洪设计、乡土物种与材料的应用、生物多样性的保护意识、材料的节约与循环利用、场地废弃物的再利用等。

1. 绸带
2.5. 水门码头
3.4. 透视
6. 入口区
7. 芦荻台

3

4

5

6

7

Shanghai needs to develop greater areas of grass, forest and wet lands, according to an overall plan of the city authorities. One of the major studies is to protect and make use of a wetland that is located inside a reserved land for Shanghai Expo. Under the theme of Better City, Better Life of Shanghai Expo, the design is based on connecting and integrating the man-made and naturally developed wetlands.

DESIGN CHARACTER-4 KINDS OF CIVILIZATIONS CHAINED BY A BLUE RIBBON

Blue ribbon area

Riverside bulrush area

Located in the north side of the land, with walking paths crossing it.

Purification wetland area

A wetland creates in the middle of the land, leading to Huangpu River and providing scenery water for the Expo area.

Ladder field area

Based on 5 meter altitude gap of the flood bank and wetland, a ladder field is created with forest shadows.

Wetland protection area

To maintain the wetland for bird collection.

There are Air Garden plaza (a result of transformed workshops), water pier plaza and the floating garden (connected to the display area).

There are a lot of areas for experiencing people.

Sidewalk networks

A convenient sidewalk system connecting each experiencing areas.

The 4 civilizations refer to wetland, agriculture, workshop heritage and post-industry eco culture.

PROBLEMS AND CHALLENGES FACING THE DESIGN

The Houtan wetland is an only one at the side of Huangpu river. Over the past 2,000 years, it has witnessed the ups and downs of the agriculture economy in Shanghai. The developing city will make it one of memorial and experiencing place for the past and future years.

The second problem facing the designers is how to readjust and purify the water, upgrade the biodiversity level and naturalize the wetland.

The third problem facing the designers is how to disperse people visiting the Expo.

The fourth problem facing the designers is how to solving the altitude level problems between the flood bank and the wetland.

The fifth problem facing the designers is how to make a grass land for the Expo and a wetland park of the city after the Expo.

The sixth problem facing the designers is how to make use of the modern concept and advanced technologies to create such a wetland park.

6 MAJOR COUNTERMEASURES

The 4 civilizations announce the land skeleton

The project highlights the time skeleton, space background and land characters, putting it into a 4-layer experiencing place.

To create a wetland base

The project includes 4 parts to collectively finish the jobs of filtration, purification, damp proof and revetment.

A varied traffic system

The multi-lane traffic network can ensure the connection of the land and outside world as well as the access inside the land.

Ladder fields tacking the altitude levels

Located in the south end of the land, it is a transitional area of the land and the city in addition to solving the altitude level problem.

Flexible design for multi purposes

The functional buildings and public service facilities adopt more strippable and recycling materials for the transformable use after the Expo event.

Eco concept leads the technology orientation

The eco and humanity concepts are dominating the entire design process, in particular the rebuilding of the wetland park, recycling of the workshop heritage and other sustainable measures and technologies.

Landscape Design for Wetland Garden of Shenzhen OCT EAST CO.,Ltd.

深圳东部华侨城湿地花园景观设计

1. 红色方舟（湿地廊桥）整体透视效果

2. 花桥广场 整个湿地花园里的主轴线，与水广场的结合为人们游憩提供了丰富的趣味性场所

深圳东部华侨城湿地花园是东部华侨城大景区的一个组成部分，是一个融艺术与生态魅力于一体的山谷湖畔区域，它将是整个园区内最具想象度和当代艺术思想开放活力的部分。它将售卖话题、事件、主张和生态本身。它将是花的世界、绿的世界、艺术家的世界。

花桥广场

花桥广场，整个湿地花园里的主轴线，贯穿了入口广场至湿地廊桥的整个空间。

花桥广场的造型源自于花朵，一个个花瓣构筑了人流的中心节点，为人们游憩提供了丰富的趣味性场所。如花撒落地面，灵动、趣味，又充满了花信的期冀。

红色方舟（湿地廊桥）

太阳能红树——靠近红色方舟（湿地廊桥）错落有致的现代景观构筑群，湿地植物红树林的抽象化。一路走去，角度多变的枝丫，连续反复的红色带予人新的感受。并在红树顶端置放红色太阳能板进行能源收集，以形成特有的生态构筑。

红色方舟（湿地廊桥）——雕塑般的红色钢构，简洁又不失细部的装置，纵向延伸，直至水域，仿若水中方舟。

城市的迅速扩张和经济的快速发展，使生态危机出现和持续，景观设计应注重在具有一定生态现实和最具想象度的区域中对生态观念的象征性表现。深圳东部华侨城湿地花园的区域特性决定了该案的意义已经不仅局限于景区娱乐休憩的范畴之内，而在于更大意义上的生态文明建设和生态教育。方舟构筑采用警醒的红色和贮满视觉凝聚力、表达震撼力的方舟骨架般的构成语言，在水畔序列展开对生态理念的感性显现。并取用圣经《创世纪》中诺亚方舟（Noah's Ark）传说之名和造型，隐喻对生态危机和生态灾难的拯救。逐渐脱

2

1

2

离船舶写实造型和形似的追求，捕捉方舟骨架间自然光影和色彩变换的效果，形成富有变化的空间表情、丰富的自然光线和活跃的信息湿地体验空间。

把自然与文化、形式与功能、生态与娱乐真正全面的融合，形成最具想象度、动态的、可参与的户外生态博物馆和多种涵义的载体——红色方舟。可向远处眺望也可在内部向外观望，同时亦作为展示空间使用。将动物展示和各种电子信息互动使红色方舟（湿地廊桥）功能重叠多变，使游人从全新的视点观察湿地，诠释湿地。这是一次关乎生态娱乐和生态教育的设计。

1. 红色方舟的甲板区域　可作为诗歌朗诵、时装发布会、创意酒会等艺术时尚区域

2. 红色方舟（湿地廊桥）　逐渐脱离船舶写实造型和形似的追求，捕捉方舟骨架间自然光影和色彩变换的效果

3. 总平面图　深圳东部华侨城湿地花园是东部华侨城大景区的一个组成部分，是一个融艺术与生态魅力于一体的山谷湖畔区域，它将是整个园区内最具想象度和当代艺术思想开放活力的部分

4. 红色方舟（湿地廊桥）　雕塑般的红色钢构，双层半下沉式的造型，简洁又不失细部的装置，仿若水中一叶红色方舟

3

4

1

2

The Huaqiaocheng wetland garden in eastern Shenzhen is part of the Huaqiaocheng grand sceneries. A valley and lakeside area combining the art and ecology charming, it is part of the most imaginable and of modern art spirit. It is a world of flower, green color and artist.

The location of the wetland garden determines that it is not only an area for the entertainment activities, but also a place for ecology civilization and education. The designers borrow the Noah's Ark image for a metaphor of the ecological crisis and the salvation of the disaster. The Noah's Ark is painted by red to highlight the convulsion of the framework.

An ecology museum, the red ark is made for a distant view from other place and from inside to outside. It is also a display environment for visitors to look at and interpret the wetland from a fresh new angle.

FLOWER BRIDGE SQUARE

The square is the key axle of the wetland garden, leading from the entrance square to the wetland corridor. The shape is based on flower and each flower provides an interesting place for the entertainment people.

RED NOAH'S ARK (WETLAND CORRIDOR)

The modern buildings close to the corridor are abstracted mangrove. The changeable twigs and continuous red color hit people with fresh feelings. There are red solar power boards on the top of the mangrove. The facilities extend into the water like a red Ark.

1. 太阳能红树　靠近红色方舟（湿地廊桥）错落有致的现代景观构筑群，湿地生态植物红树林的抽象化

2. 红色方舟的甲板结构　可作为观景平台，游客可登高远眺湿地花园不同的风景

3~5. 红色方舟（湿地廊桥）

6. 方舟甲板上构成式的观演区域

3

4

5

6

1

2

1.2. 红色方舟（湿地廊桥）局部透视效果

3. 栈道

3

工程档案

工程名称
深圳东部华侨城湿地花园景观设计

工程地点
广东省深圳市

设计单位
广州土人景观顾问有限公司

设计师
庞伟、张健、陈瑜璟、李津、魏敏、黄志坚
朱开平、谭观朱、阮柳植、张桂平、陈彩霞、
胡电智、卫冠杰、房丹凤

项目景观总占地
22.8公顷

设计完成时间
2007年2月

竣工时间
2007年9月

Landscape design of the Peach Park, Yongdingmenwai, Beijing

北京永定门外桃园公园景观设计

永定门外桃园公园位于北京市崇文区永外大街东侧辅路路南，这里作为老街巷、老建筑拆迁后形成的一片公共绿地，对城市脉络的延续和对现有资源的利用成为我们设计中的重点。

如同全国各地不断建设中的地域一样，这里旧有的建设于不同时期的不同风格的建筑看似混乱，却被原住民认同接受，成为生活中不可或缺的部分。地块中拆除周围的平房后，部分房屋基础高出道路将近1米，形成"台"字形，"台"上有三棵千头椿树，原来应该是大院里的庭荫树，多年后已长得郁郁葱葱。因为这三棵树我们保留了这部分基址和高台，并依照原有的建筑空间设置了休息的场地，使这里的老居民老街坊依旧能够背靠大树，沐浴着暖暖的阳光，懒懒地遥望永定门。原有低陷的街道所形成的谷地空间也被保留，两侧种上碧桃、海棠等蔷薇科植物，塑造成为视线能通达永定门的"桃花谷"。

对墙体的设计一直是我们关注的焦点。城墙脚、皇城根、胡同内的青墙灰瓦，"墙"几乎成为京城老百姓生活中最强烈的感受，也成为一种京城文化的精神寄托。而我们运用现代技术来重新诠释一个古老的题材，借助计算机重新定义墙体，运用抽象的空间组合方式表达传统文化的隐喻，摆脱了单纯符号的苍白无力。院、墙、竹、台阶这些意念中与传统文化意境紧密结合的事物成为我们表达的语汇，将园林环境与人们的怀旧情结自然地联结在一起。

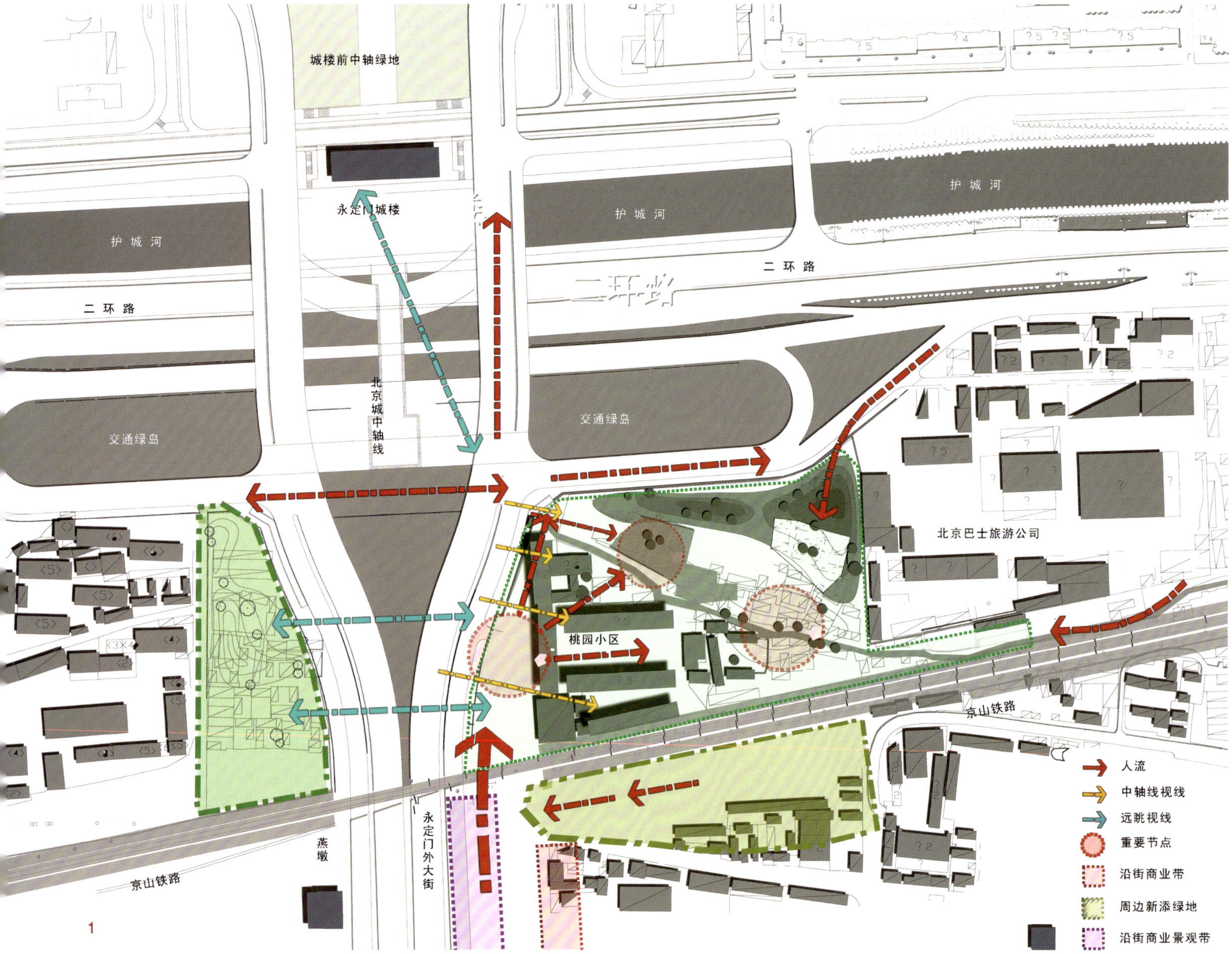

1

此项目的最终设计保留了许多大树，但都没有准确的定位，为了使院落与大树能自然地结合，设计时采用了"因树构景"，用院落构建大树之间的联系。大树本来就是原有大杂院中的一部分，树之间的尺度正是原有院落的尺度，这样一来反而与原建筑庭院的布局不谋而合，产生如故地重游般的归属感。

在材料的运用上，我们也采用了白描的手法，用实木、砖、石料等构筑庭院，使材料本身表达朴实易懂的意境。我们采用了规格大小与青砖相同的混凝土砖来代替原来的青砖，但因为混凝土材料特性与烧制的砖不同，它致密、色差小、易变灰，因此在色彩上选择了较深的咖啡色与较浅的咖啡色，以期在混砌时能体现出色差的变化。混凝土的强度使冬季冻涨后砖体不会发生脆裂，有效地弥补了在北方青砖铺底的缺陷。

工程档案

工程名称
北京永定门外桃园公园景观设计

工程地点
北京市崇文区

设计单位
北京北林地景园林规划设计院

设计师
张璐、陈立基、王斌、张亦箭、项飞、石丽平、马亚培

项目景观总占地
1.9公顷

单位平米造价
150元

设计时间
2006年7月

竣工时间
2007年6月

1. 规划视线及人流分析图
2.4. 花园一角
3. 中心花园实景
5. 广场一角

2

3

4

5

1

2

3

4

5

The Peach Park is located in the south end of the east side service road of Yongwai street of Chongwen Dist, Beijing. A public land for greening as a result of the relocation of veteran blocks and buildings, it needs to highlight continuing venation of the city and making use of the current sources.

Similar to situations in all over China, buildings in this area of varied times are acceptable for the local residents as an irrevocable part of their life. Partial house pedestals are nearly 1 meter higher than the surrounding areas and there are three Chinese toons which are of a wild profusion of vegetation over the years. Thanks to these three trees, the pedestals were reserved and a leisure site was arranged, allowing people there view the Yongdingmen, leaning against the trees under sun shines. The lower part of the previous street is reserved too, with flowering peach and cherry-apple trees being planted alongside both sides, forming a "peach blossom volley".

Much attention has been attached to the design of the walls. Beijing residents have been deeply impressed by the dark wall and grey tiles all over the city. The walls hit the residents most directly and carry out the capital culture in an emotional manner. Modern technologies have been employed to interpret the traditional cultural metaphor, supported by computer and abstract rearrangements of the space, free from the monotonousness of the individual factors. Compound, wall, bamboo and step, which carried a lot of images in the traditional culture, are used to naturally link the landscape environment with the reminiscence complex of people over the past years.

A lot of trees are remained, but lack precise positioning. To better combine the compounds and trees, a tree-oriented idea was adopted and the compounds act to link the trees. The trees, as part of the compounds, measure the sizes among the previous compounds. The results are compatible with the previous structures of the compounds, reminding people of their memories on the sites.

Concerning materials, wood, brick and stone are used in a natural manner, to let materials tell the stories. Concrete bricks of the same size were used as well. But these bricks are different in terms of density, color difference and they are easy to become greyer. Therefore, dark and light coffee colors have been chosen to show the differences in combined use. Concrete bricks will not crack in cold winter as normal bricks do, a headache in north China's construction industry.

1. 中心游园效果图
2. 中心游园
3. 园路
4. 入口广场实景
5. 中心花园实景
6. 鸟瞰图

6

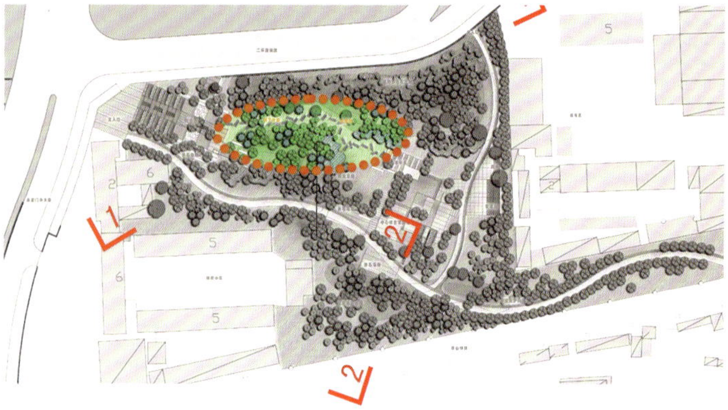

2-2断面示意

铁路边缘绿化隔离带采用加密种植的方式结合地形，营造良好的屏蔽噪音与遮挡视线的隔离带。同时在靠近园区的部分种植姿态优美及可观花观叶的树种，增加隔离带的实用性，可观赏性。

利用基址中原有地形营造一种自然山谷的感觉，并遍种蔷薇科植物，当春风拂过京城大地，人们透过这片山花烂漫之谷欣赏，回味永定门散发出的那百年古都的魅力……

1-1断面示意

1

2

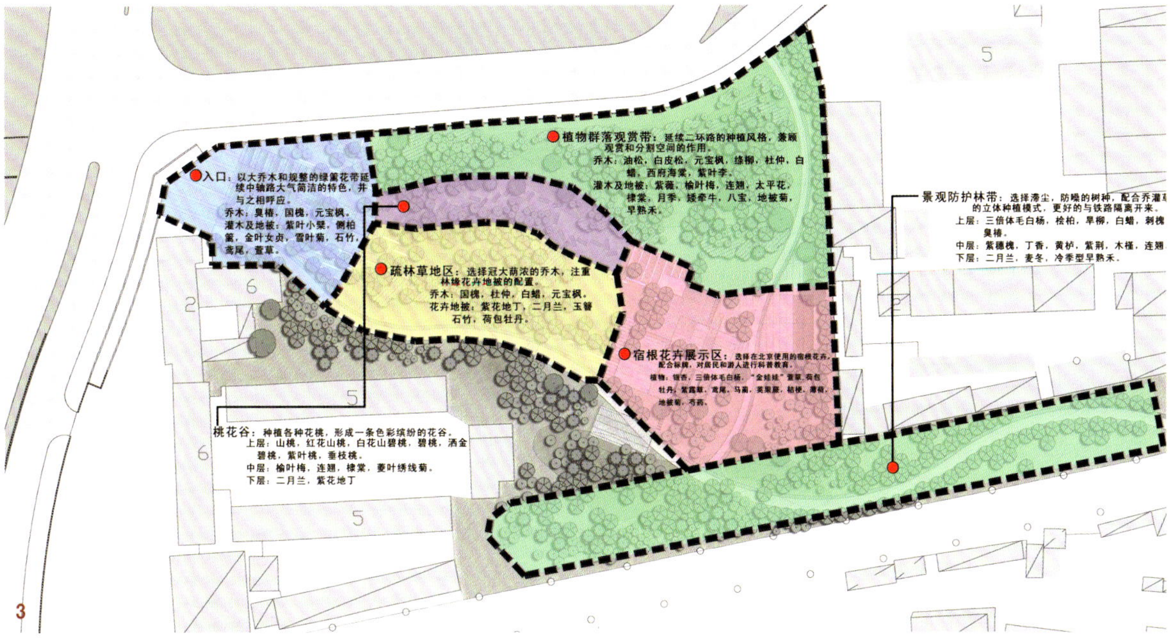

3

1. 桃谷节点示意图
2. 北区界面效果图
3. 种植图
4. 总平面图
5.6. 中心花园实景

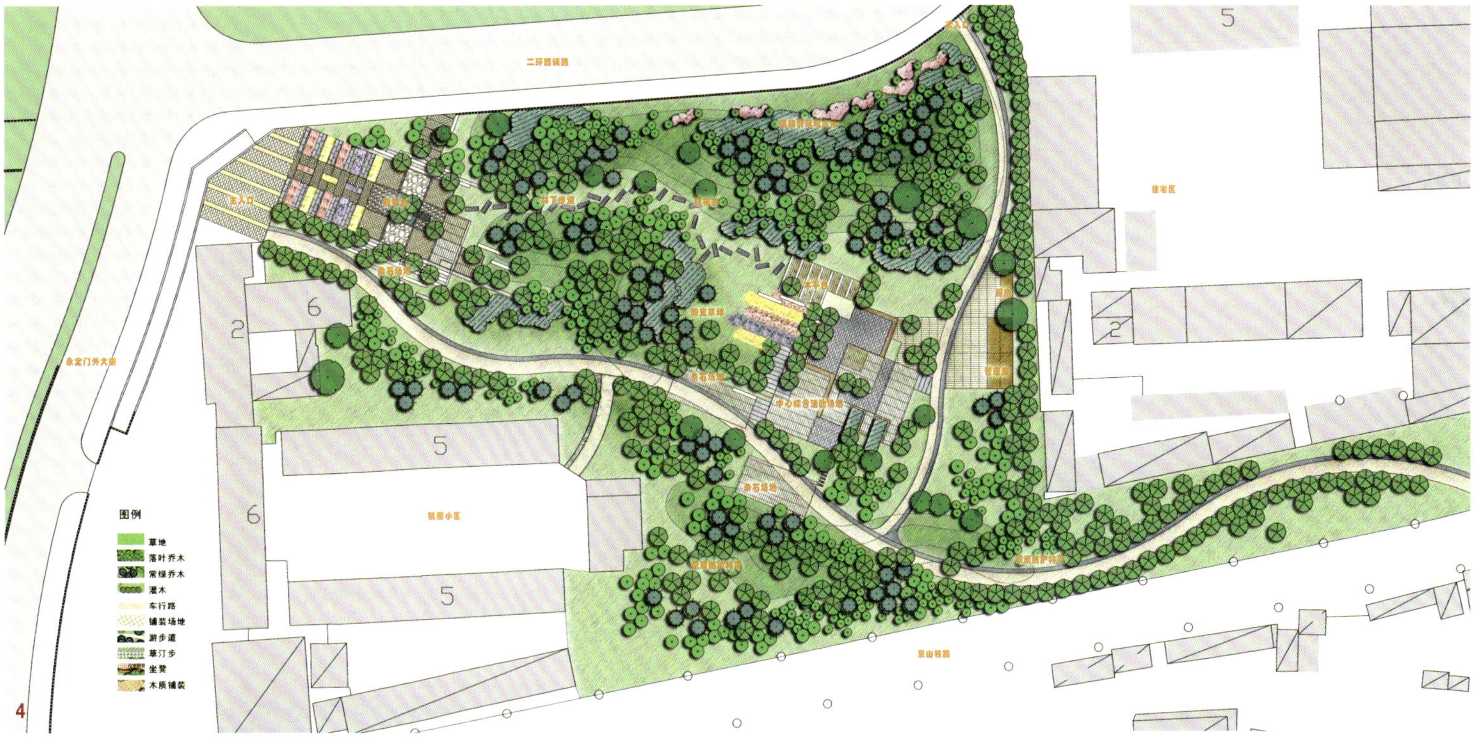

4

5

6

Landscape design of Yandun heritage Park, Beijing

北京燕墩遗址公园景观设计

设计理念

本项目位于北京城市中轴线上的永定门西南侧，京山铁路的北侧，用地方式为道路绿化，街心绿地为永定门南中轴线城区景观起点，是北京中轴线的南起点，地理位置重要。项目用地北望永定门城楼，南侧为始建于元代的燕墩遗址。历史上，这里是京城南大门的交通要塞，京杭运河漕运的北终点。如今，在燕墩遗址公园这个开放的城市公共绿地中，回望历史、感受今天、享受生活；展现并回顾京城悠远的历史和文化，将历史与现代自然融合；为周边市民提供日常的户外休闲娱乐场所成为此次设计的基本立意。

设计手法

在设计中，充分强调了"望城"的概念，以用地地形和原生大树为基组织景点，让人可以南看燕墩、北望城门，回顾历史，并通过对植物和地形的组织，形成远近看城的效果，产生多层次、递进的历史认识。利用借景和障景的手法，形成不同距离、不同高度、远近有别的看点，形成多重的对一南一北两个建筑的印象，从而深刻体味历史、文化的悠远。

核心的景点利用了原有的古槐、组织凉亭和台地广场，形成了远近不同的看城场地，同时成为居民的户外休闲场所；在靠近平台的西侧，设置了两个景观雕塑，形成"望城"的取景框效果。

地形组织

利用原有场地高差，随地就势，因高就低，配合成片栽植的乡土树林，形成起伏变化、层次错落的园林景观。

植物种植

延续南中轴整体景观特征，以乡土树种大片种植成林，同时通过花灌木和地被植物的组织处理，充分体现出四季的变化特征。主要苗木品种为银杏、白蜡、玉兰、云杉、油松、桧柏，灌木品种为西府海棠、紫叶碧桃、红王子锦带、丛生紫薇等，地被品种为丰花月季、德国景天、蓝羊茅、万寿菜等。

交通组织

项目位于永定门外西侧，兼顾以绿植为主的前提下，增加活动停留、休息的场所。主干道南北向园路为主要景观步道，穿过了园内的7个主要景观节点，与西侧的市民住宅及东侧步行道有横向的道路连接，形成了开放式的绿地花园。通过地形的开合，植被层次的错落，塑造宜开宜闭的视线变化。主要景观节点为北入口区、中央区域、四眺台、荫下平台、雕塑式亭子及管理用房。材料、造型、体量采用了与环境相应的灰砖，营造城台花园的意境，城墙与主要乔木种植的向心方向为北侧的永定门城楼，充分体现"望城"的立意。

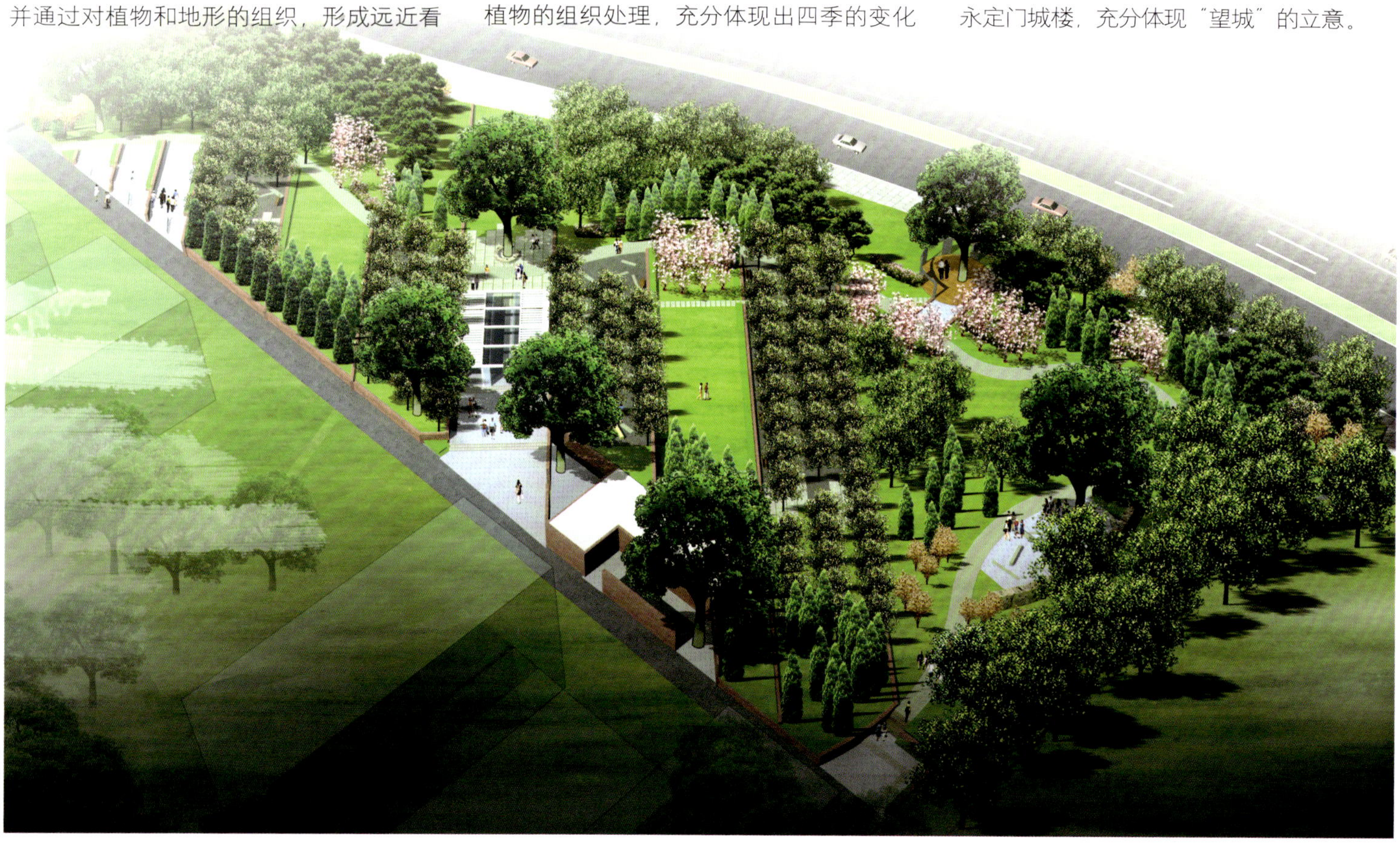

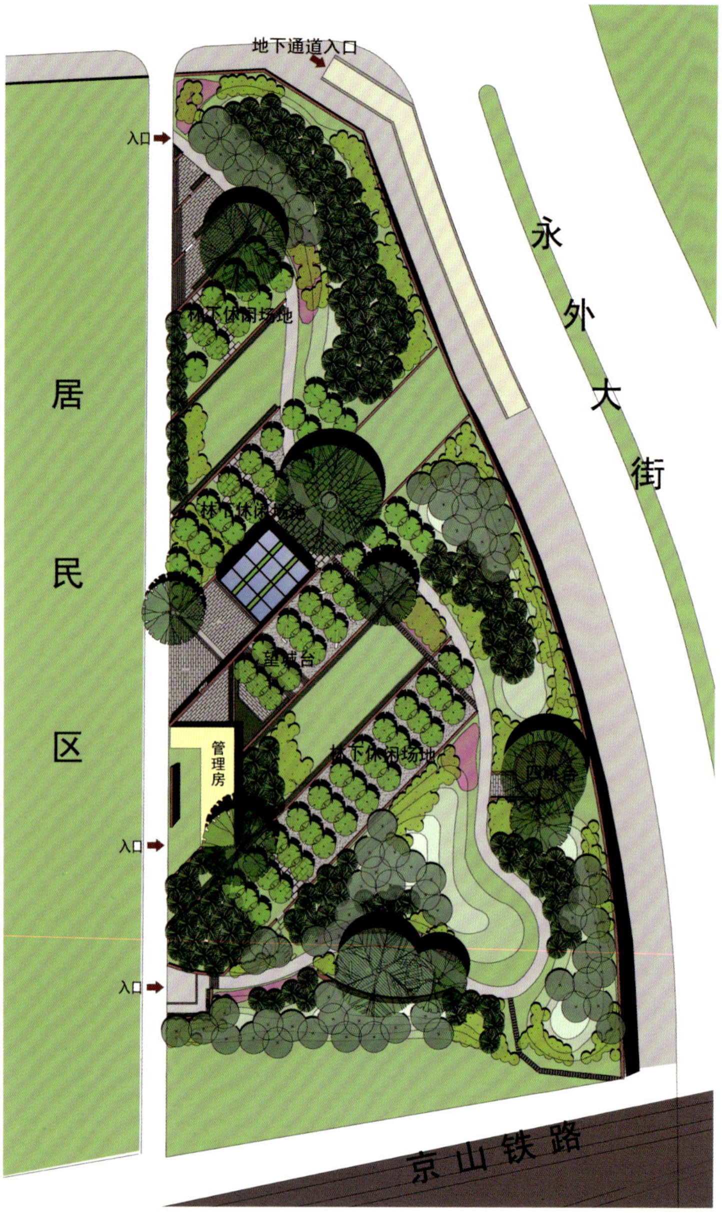

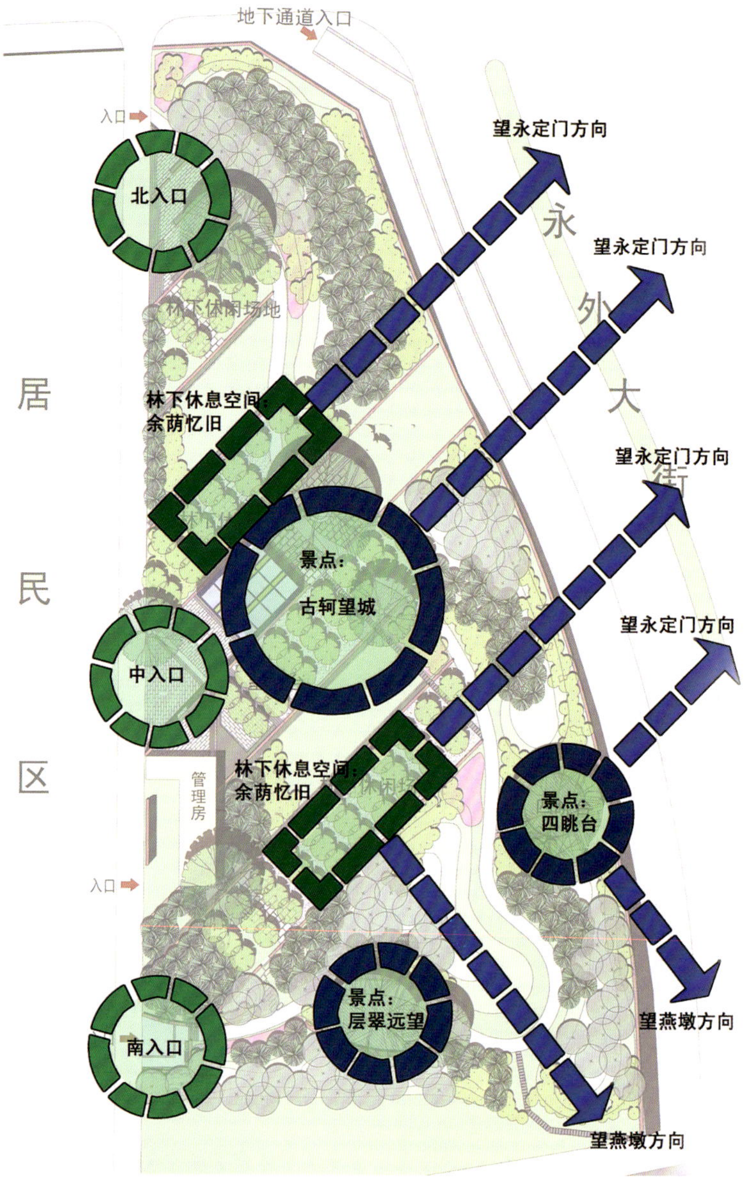

工程档案

工程名称
北京燕墩遗址公园景观设计

工程地点
北京市崇文区

设计单位
优地联合（北京）建筑景观设计咨询有限公司

设计师
李健宏、秦秀娥、严骏、王欣妍、冯丽、陈雷、焦燕菁

项目景观总占地
1.1公顷

单位平米造价
200元

设计完成时间
2006年8月

竣工时间
2006年10月

DESIGN CONCEPT

The project is located in the southwest side of the Yongdingmen and north side of the Beijing-Shanhaiguan Railway line. The nature of the land use is greening and the street centre greening land is the south end starting line of the Beijing Central Line. The project land is surrounded by the Yongdingmen city tower in the north end and the Yandun Heritage Site founded in the Yuan dynasty. Historically, the area is key to the south part of Beijing and it was the northern end of the Beijing Hangzhou Canal. Now the Yandun Heritage Park acts to integrate the history and modern life and a park for entertainment and sports for people living around it.

DESIGN APPROACHES

The design has managed to allow people to view the Yandun and city gate, and the plants and landform help view the city historically.

The core landscapes make use of the existing ancient pagoda tree, arbor and platform plaza, allowing people to watch the city from various distances. There are two scenery sculptures at the west side of the arbor.

LANDFORM ARRANGEMENT

Compatible with the existing land, local plants were planted to form a scenery landscape which are changeable and stand at various levels.

PLANTS

A lot of local plants were planted to show the changes in the 4 seasons.

PATH ARRANGEMENT

The Xiangyuan Road is main path, connecting 7 main sceneries inside the park. It is also connected to the west side residential area and the walking path in the east side by transverse paths. Gray bricks were used to remind people the history of Beijing.

Planning design of Huantai culture and sports Park, Zibo city

淄博桓台县文体公园规划设计

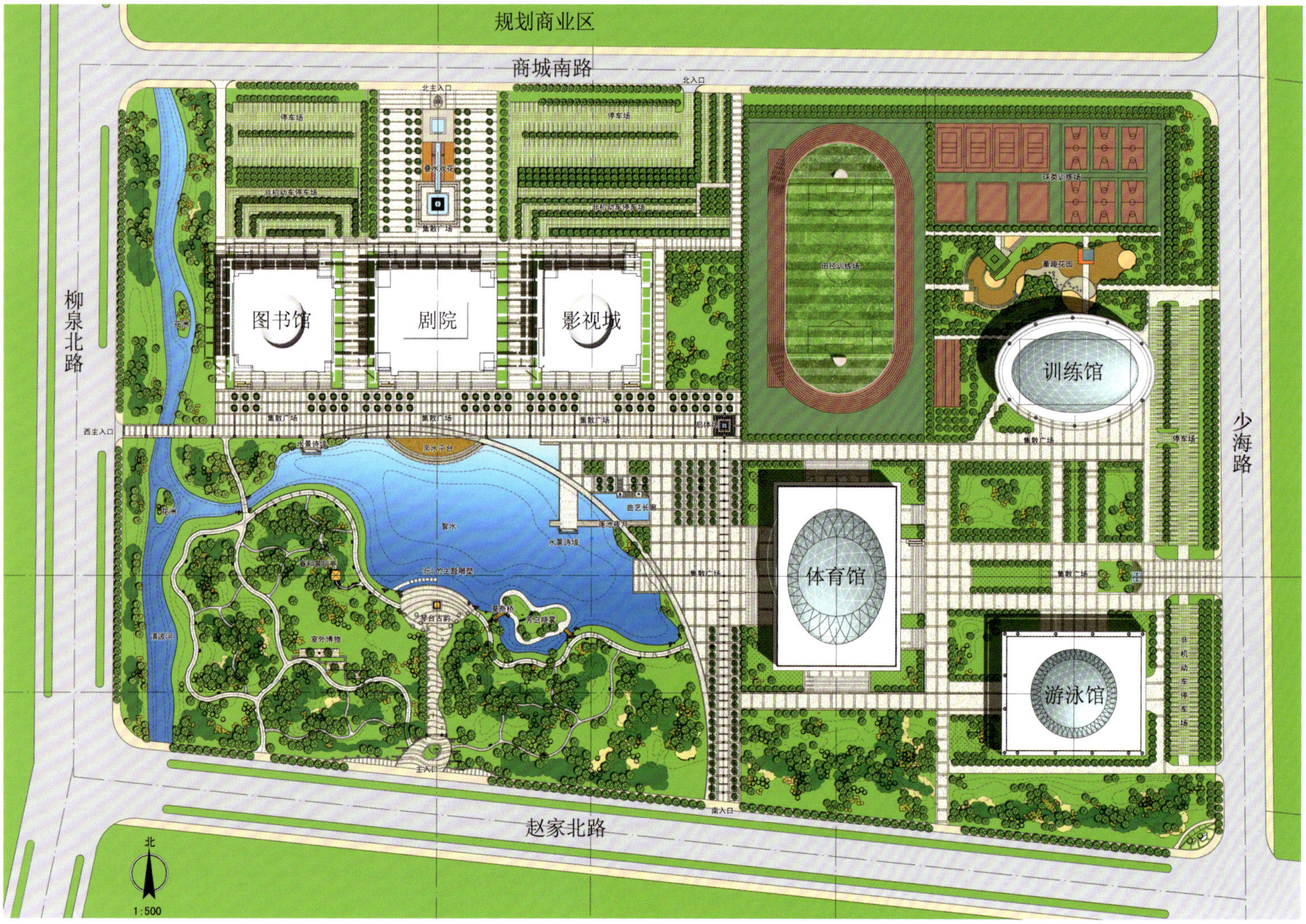
规划商业区
商城南路
图书馆
剧院
影视城
训练馆
体育馆
游泳馆
柳泉北路
少海路
赵家北路
北
1:500

1. 莲池月夜
2. 大门
3. 曲径
4. 拱桥

概况

桓台文体公园位于桓台县中心区，规划面积约33.8公顷，基址现状地势平坦，西侧有一条水系穿越园区，园区内的空间布局紧紧围绕着文化中心与体育中心两大组团功能建筑群展开，对其如何把握与呼应是本次规划设计的核心内容与关键点。

本方案力求将桓台文体公园建设成为与现代城市生活相适应、以鲜明的地域文化和体育文化内涵为特色、以休闲娱乐、康体健身为主要功能的开放式城市公园。

地域文化内涵和体育文化内涵的确定

地域文化内涵：桓台地区具有悠久的历史，千百年来形成了以齐文化为代表的丰厚文化积淀，把握住中华文化的内核，就是对“齐文化”核心的最好体现。

文化内涵融入公园设计中，应采用分层深入的方式。在公园山水体系的布局方面，突出以自然为主、恢弘大气的整体风格，同时，在公园总体形态方面，不搞单一中心主体型公园结构，而是采用分中心、多内核的多中心式公园结构，在公园内形成风格不同、特征各异的区域景观。这不但与桓台文体休闲公园的场地特征相符合，同时也是“海纳百川”文化核心的具体体现。

体育文化内涵：体育运动是重要的体育文化内容，它所代表的不仅仅是运动竞技比赛，更有着丰富的文化内涵和教育意义。对人们进行体育教育将有助于提高他们的身体素质，形成正确的体育锻炼意识，达到全民健身的目的。

公园总体构思与结构的确定

公园总体构思

桓台文体公园的文化中心与体育中心建筑群占地面积3.4公顷，我们设计的目的是把建筑融入自然山水之中，将公园设计成为以自然形态为主体，能够为市民提供休闲娱乐、健身活动的现代城市公园。解决好自然与建筑的关系、文化中心及体育场馆的人流集散场地与公园绿地空间的关系，是本次设计的首要任务。

文化、体育主题是本方案的特色，这

1

BRIEF INTRODUCTION

Huantai Park is located in the centre of the town. The 33.8 hectare land is flat, with a river passing at the western side. The space arrangements inside the garden stick to a culture centre and sports centre and it is the key of the design.

The program expects to build a garden compatible with modern life and being with distinctive local characters.

DETERMINATIONS OF THE CULTURE AND SPORTS CONNOTATIONS

Culture connotations: the town has a 1,000 years history, based on the so-called Qi Culture.The culture connotations are introduced by varied layers. The river systems emphasize nature and wildness and concerning the overall forms, multi centres and multi cores are introduced. The sceneries are different from each others.

Sports culture connotations. The garden help people realize the importance of sports.

DETERMINATIONS OF THE OVERALL CONCEPTION AND STRUCTURE

Overall conception

The culture and sports centre has a floor space of 3.4 hectare. The buildings are expected to be harmony with the environment. The design needs to tackle the relationships between plaza and greening lands as well.

The culture and sports themes demand Happiness, which means both optimistic and music. The design is expected to create a happy ground by means of garden, building and the arts.

Structural characters

It refers to the transitional area from the plaza to the greening garden.

2

3

4

5

1. 室外展廊
2. 大门
3. 湿岛探幽
4. 山光水色
5. 北广场

Landscape design of Chanhe riverside Park, Xi'an

西安浐河滨河公园景观设计

滨河公园作为水岸东方项目配套的社区公园，是西安城东城市景观的重要组成部分。它东边跨市政道路临水岸的东方小区，西边为橡胶坝拦截的浐河，为一个狭长的临水公园，这个项目主要服务于本社区的居民，同时也对社会开放，它是联系城东与市区的主要节点，也是西安的东大门，展示了西安的新形象。

设计思路

项目本着"母亲河畔，都市水岸生活"的综合发展思路，吸收当地的人文、自然景观特色，坚持创新的理念，创建出新景观、新概念、新思想。通过对公园的整体规划设计，确定中心公园的地位和作用，同时使公园各项设施布局合理、各种功能逐步完善，使公园和城东周边环境形成一个和谐统一的整体，营造一个景观优美、环境自然亲切、具有地方特色和文化内涵的现代化城市中心公园。设计布局中把传统文化穿插布置在各个景点中，以小品、雕塑等形式出现。

总体布局

以水为中心，采用功能中心（入口景观区、景观休闲区、主入口及中心结点）分散和四条路（城市车行景观大道、公园中的小路、河岸边大道、河滨栈道）贯穿的总体布局，各个景观节点采用散点布置，每个节点一个中心主题，互相呼应，互相联系。

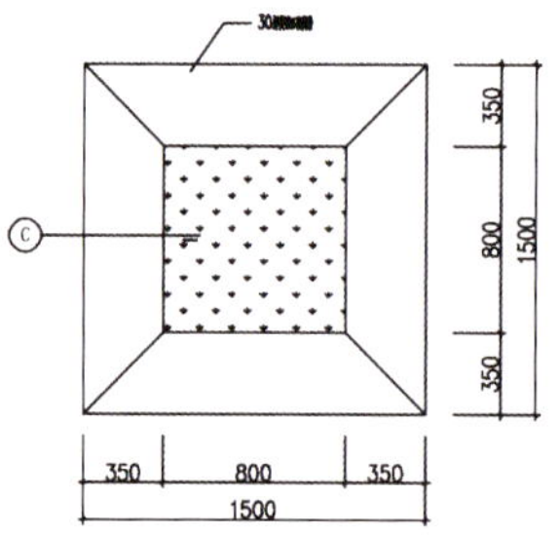

A 方形树池平面图 比例1:20

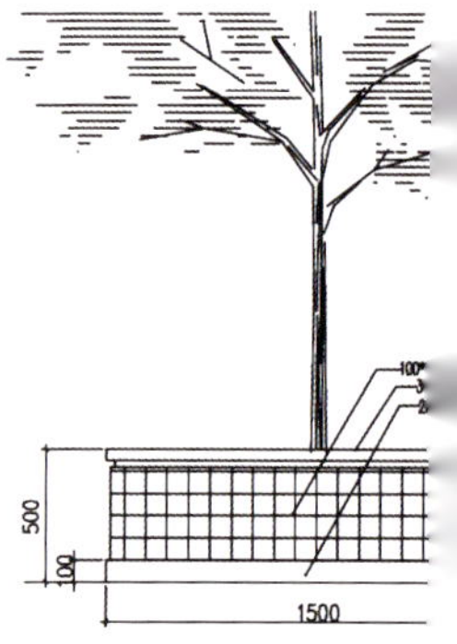

B 方形树池立面图

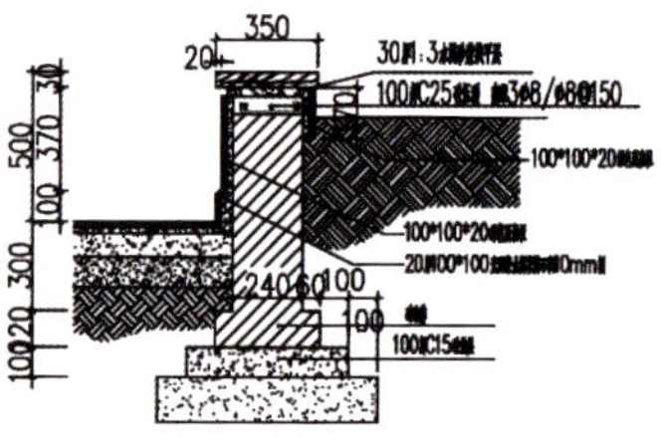

C 方形树池剖面图 比例1:20

总体规划设计一

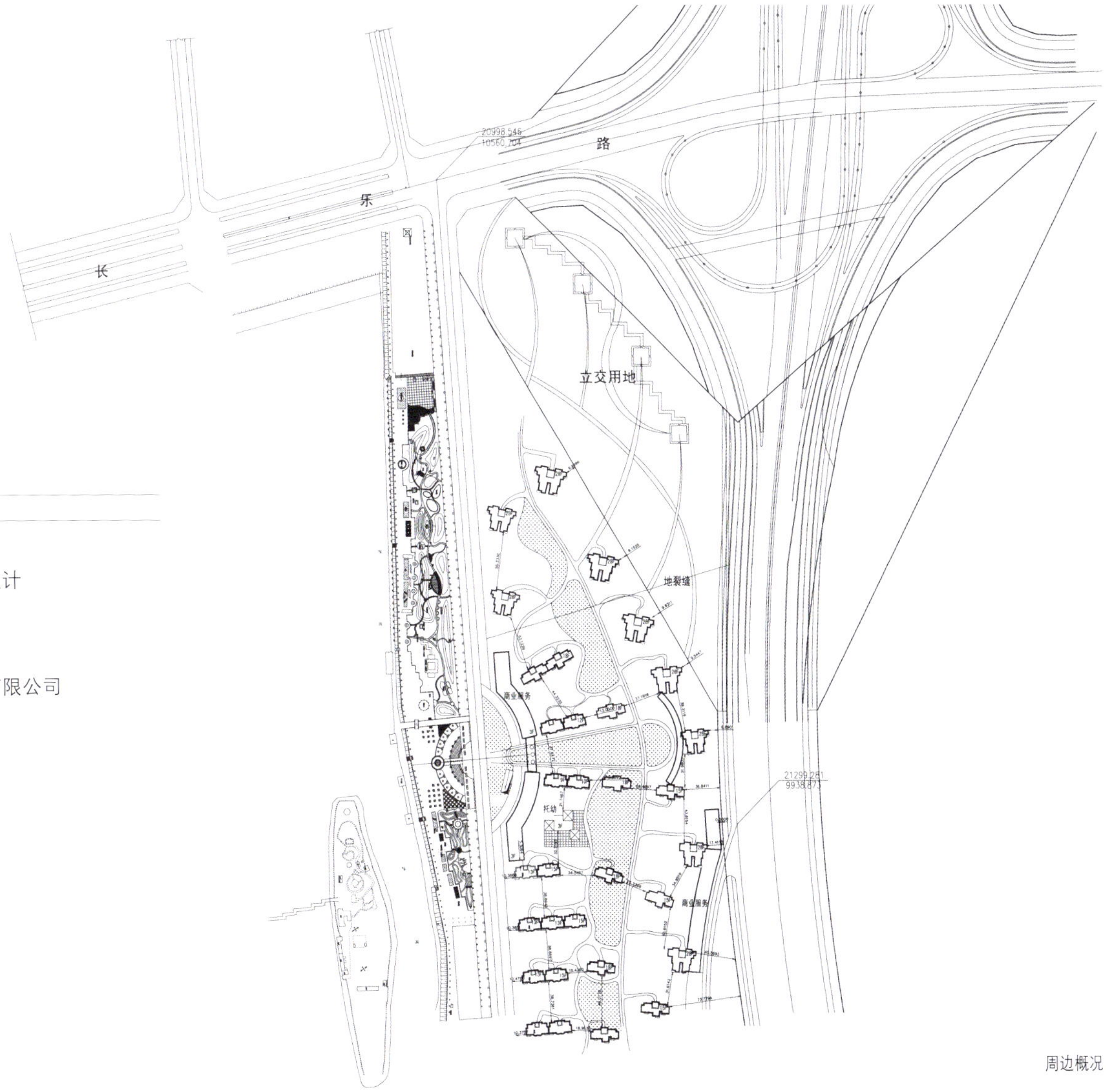

周边概况

工程档案

工程名称
西安浐河滨河公园景观设计

工程地点
陕西省西安市

设计单位
陕西广跃建设实业发展有限公司

设计师
田春生

项目景观总占地
13.73公顷

单位平米造价
230元

设计时间
2007年3月

竣工时间
在建

1

2

道路系统规划

城市主干道为整个社区和滨河花园的主干道路，用优美的行道树及景观设施，构成一个美丽的景观大道，电瓶车、双人自行车均可通行；滨河路是观赏河岸风景的主道路，设置柳树及坐椅；河边亲水平台路是最靠近水的路，采取多样化的铺装；绿地中间的园路是为游人欣赏风景、体验生态所布置的道路，做成多样化的铺装，形成宽为1～1.5米的丛林步道。

植物配置规划

根据当地的气候特点，植物群落是以常绿针叶树为主与落叶阔叶林混交出现，充分考虑冬日对阳光的需求，大量运用的植物种植群落应具有以下层次结构：上层大乔木以落叶阔叶树为主，形成上层界面空间，以保证夏季的浓荫与冬季充足的阳光；中层乔灌木以常绿针叶树种为主，同时结合观花、叶、果、杆及芳香物种，形成主要植物景观感受界面空间；下层是耐荫的低矮花灌木、地被及缀花草地。

照明设计

在园内各处分别布置各类景观照明灯，既满足夜间照明需要，又有装饰效果，各处草地上皆置以草坪灯，在台阶下藏有灯带。在大树底下设计有投射灯,岸边沿河有霓虹灯、人行道路灯、岸边景观灯、草坪灯、大树下照树灯、壁灯等。

1.2. 总体规划设计
3. 曲水荷风景点
4. 水边草亭夜景
5. 曲桥夜景
6.7. 广场夜景
8. 景观灯柱夜景
9. 绿地夜景
10. 河岸夜景
11. 水边咖啡屋夜景

3

4

5

6

7

8

9

10

11

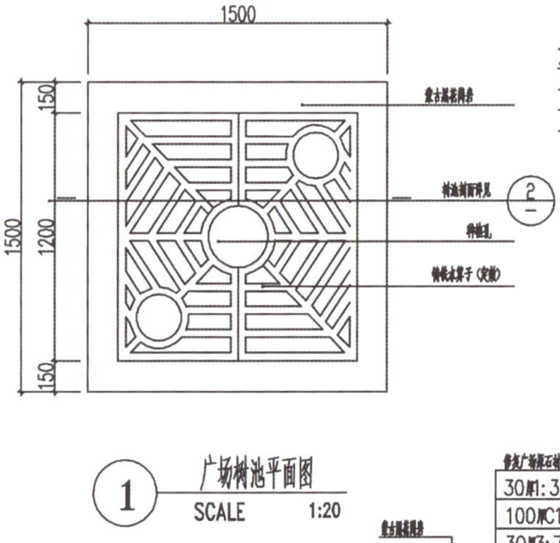

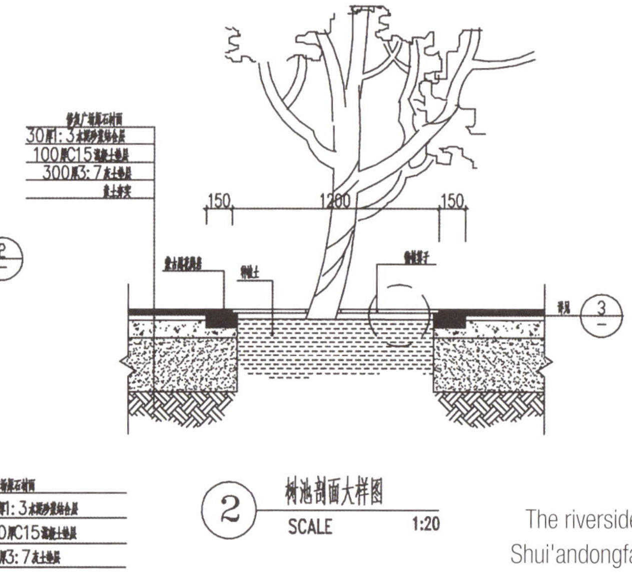

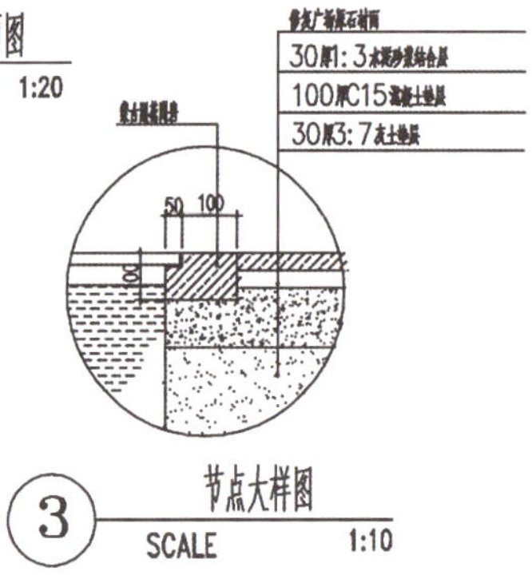

1. 水岸绿地广场夜景
2. 中心节点广场
3. 南入口大型水幕雕塑

The riverside park, part of a community park of the Shui'andongfang project, is important for the sceneries of east Xi'an city. A narrow park, with immediatc limits in its east and west ends, it mainly serves local residents and people from outside as well. It is a main point connecting the east part and the city and serves to display the new image of Xi'an.

DESIGN CONCEPTION

The project highlights new scenery, new concept and new idea, making itself and the environment a harmonious complex. The traditional culture is showed by ornaments and sculptures dropped among each site.

OVERALL LAYOUT

Based on water, the decentralized functional areas are connected by 4 roads. Each point has its own theme and they echo to one another.

ROAD SYSTEM PLANNING

The main roads are decorated by street trees and sceneries for storage battery cars and bicycles. Alongside the riverside roads are willows and chairs. The riverside podiums are decorated by colorful floors. There are1~ 1.5 meters wide paths across the grass land.

PLANTS ARRANGEMENT PLAN

The plants are planted for the sakes of sunshine, level space and scenery feelings.

LIGHTING DESIGN

All lighting lamps serve at night and as decoration. There are grass lamps on every grass land and lighting belts under the steps. There are a lot of other kinds of lamps in other places.

2

3

Environment improvement around the Potala Palace and planning design of Zongjiaolukang Park rebuild project

拉萨布达拉宫周边环境整治及宗角禄康公园改造规划设计

场地概述

布达拉宫坐落在拉萨市风景秀丽的红山上。宗角禄康公园成U字形从东、西、北三个方向环抱布达拉宫。本项目基地原有的状况存在以下问题：主题内容不多，缺少地方特色；功能分区不明确；空间层次感不强，全园同一基调；大量建筑占据公园外围，公园与城市的渗透性和通透性较差；园路分级不明确，未成体系；水体驳岸处理简单，亲水性差，滨水休闲设施不足，防水设施不牢。植被树种单一（以藏川杨和柳树为主），缺乏层次，乔木灌木比例不协调，软质空间围合感不强，有些树木已成危树；地形简单，全园基本无地形起伏，缺乏层次；设施小品陈旧，布局不合理，缺少必要的休闲设施。

景观设计理念

城市和谐发展，城市文物的保护和城市旅游的发展是本次公园改造规划的三个出发点，因此，我们确定的公园改造规划理念为："有机生长的城市生态客厅。"作为布达拉宫后花园的宗角禄康公园，它承担着拉萨市民休闲娱乐的功能，承担疏散和接待布达拉宫游客的功能，还承担着联系拉鲁湿地和拉萨河生态链的功能。因此，本次公园改造规划要把它打造成生态与人文和谐共生的城市休闲公园。

景观设计目标

我们根据拉萨市的经济发展现状，有计划、有步骤地疏理、整治、更新布达拉宫周边的环境，使其充分反映历史风貌，同时构建新的辐射整个城市的绿色生态系统，完善康体休闲系统，给城市的生态及人居环境带来巨大的生态效益与社会效益。

2

3

景观设计策略

基于对布达拉宫周边现状和拉萨市经济发展现状的理性思考，我们提出对布达拉宫周边环境整治分“二期三步”来实现。整个整治工作分为一期改造和二期改造。

一期改造

主要是宗角禄康公园和布达拉宫转经道周边环境的改造，拆除公园内遮挡视线的建筑，恢复绿化、改造水系、整合园路、构建新景区新景点，梳理转经道及转经市场，拓宽部分转经道，改为林下转经市场。

二期改造将分为三个步骤进行

第一步：拆除自治区有线电视台职工住宅楼、沿林廊西路建筑、邮电局办公楼及邮电酒店，打开公园视线，恢复绿地，修复历史上部分水系，组织园路交通，开辟停车场。

第二步：改造西藏电视台及自治区广播电影电视局办公楼，将办公功能置换为旅游市场功能。拆除自治区组织部、民航局和邮电局南北向建筑，开通东西向视线，整合建筑庭院并将功能置换为旅游住宿和旅游餐饮功能。

第三步：根据布达拉宫文物保护要求，建设控制地带内建筑制高点为6m。故将第二步骤保留的建筑全部拆除，只留部分6m以内的建筑，并重新梳理成传统院落形式。这部分建筑将作为旅游市场、公园附属配套建筑和旅游特色餐厅。恢复绿地和历史上的水系，组织园路交通，开辟停车场。

通过“二期三步”的实施，逐步实现“有机生长的城市生态客厅”的设计理念。

1. 雾泉广场
2. 观演广场
3. 布达拉宫东侧围墙处林荫广场
4. 湖区
5. 公园入口广场
6. 下沉广场
7. 水边木质广场

第100页图注

1. 水边广场及茶室
2. 湖区
3. 七曲桥
4. 游船码头
5. 观演广场

4

5

6

7

林 廊 北 路
林 廊 西 路
北 京 中 路
娘 热 南 路
生态水景区
民俗活动区
民族风情区
龙潭胜迹区
亲水活动区
文化活动区
地域风情区
图例
转经道
保护建筑
保留改造建筑
水体
园路
广场
植被
地形
停车场
自然荒石
水中岛
N
0 10 30 60m

1

2

3

4

5

LAND BRIEFING

The Potala Palace is located on the Red Hill of Lhasa and it is hugged by U shaped Zongjiaolukang garden from the east, west and north sides. Current problems: a lack of contents and local touches; unclear functioning; weak space arrangement; too many buildings around garden; unclassified garden path; a lack of waterside facilities; single plants; flat land; aged and disordered ornaments.

SCENERY DESIGN CONCEPT

The objectives of the redesign this time include a harmonious city, relic protection and city tourism, making the garden "a dynamic eco living room". A backyard garden of the Potala Palace, the Zongjiaolukang garden was designed for resident's entertainment, receive the Palace tourists and connect the Lalu Wetland and the Lhasa River.

SCENERY DESIGN OBJECTIVE

The improvement will phase in to benefit the entire city.

SCENERY DESIGN APPROACHES

Two periods are planned to finish the entire trimming job.

First phase. It refers to the improvement of the Zongjiaolukang garden and the praying paths around the Potala Palace, dismantle some keeping out buildings inside, greening, rebuild the rivers, integrate park paths, build new sites, and widen some praying paths.

The second phase improvement will be done by 3 steps.

First step. To dismantle some buildings, recover the greening land and build a parking lot.

Second step. To rebuild some buildings, making them for tourism and catering uses.

Third step. To maintain buildings lower than 6 meters in height around the Palace and make the buildings tourist market and rebuild the paths inside park.

These steps are expected to fulfill the above-mentioned design concept.

6. 文化活动广场
7. 林下小广场
8. 七曲桥

6

7

8

春游浩荡
是年年寒食
梨花时节
白锦无纹香烂漫
玉树琼苞堆雪
静夜沉沉
浮光霭霭
冷浸溶溶月
人间天上
烂银霞照通彻
浑似姑射真人
天姿灵秀
意气殊高洁
万蕊参差谁信道
不与群芳同列
浩气清英
仙才卓荦
下土难分别
瑶台归去
洞天方看清绝

工程档案

工程名称
布达拉宫周边环境整治及宗角禄康公园改造规划设计

工程地点
西藏自治区拉萨市

设计单位
中国建筑设计研究院环艺院景观所

设计师
李存东、史丽秀、赵文斌、温亚玲、史莹芳
雷洪强、巩磊、王洪涛

项目景观总占地
36.5公顷

单位平米造价
150元

设计完成时间
2006年6月

竣工时间
2006年11月

Landscape design of comprehensive facilities for disabled and Special Olympics training Base, Shanghai

上海残疾人综合设施暨特奥竞赛训练基地景观设计

基于该项目服务人群的特殊性以及该工程对上海的重要性，结合2007年特奥会的精神，设计师对项目的设计进行了特殊的定位，提出"无障碍"景观的设计概念，主要体现为"情感沟通无障碍、交通流线无障碍、身体康复无障碍、心灵康复无障碍"四个层面，以期营造出"关爱、平等、自强、和谐"的景观空间。

关爱的景观

设计师希望将"关爱"的情感融入设计创作当中，以求设计能体现社会各群体对残障人士的关爱。因此，根据使用空间的不同，设计师营造了集中交流区和散点交流区。集中交流区主要位于中央景区、培训中心、康复中心等人员集中区，提供室外课堂教学和病人之间的集体交流空间；而散点交流区则自然地穿插在庭院景观、沿河景观带等休闲区域，为病人游园赏景时提供驻足交流的空间。在整个户外环境中安排了大量的室外座椅，并很好地结合了地形与自然环境，既成为了一种系统的小品设施又构建了人与人交流休息的平台。

平等的景观

"平等的景观"就是希望不论健康人还是残障人，都能无障碍地享受同等的景观空间。首先，基地内的车行系统和人行系统都是相对独立的，人行系统的整体标高高于车行系统，这样能更好地满足安全的需要。其次，盲道、语音提示、盲文地图等组成的盲人步行系统能使盲人到达基地内正常人所能到达的任何地方。另外，设计师依照场地、造价、使用情况等多方面因素，对室内外的高差问题做了综合的调整，以1:50为最佳坡度、1:20为基本坡度的原则对各幢建筑的出入口进行了控制。

1. 会议中心外观
2. 中央景区
3. 安养楼庭院
4. 几何地形
5. 风雨廊
6. 会议中心大楼

1

2

1

2

3

自强的景观

残障人员要重拾信心，恢复身心健康，就一定要有自强不息的精神，坚持进行训练，才能更快地康复。在雨露广场和风铃广场两处设有回音壁、传声管、风铃阵等小品，可进行趣味听力训练；雨露广场上的景墙立面进行了特别的彩灯装饰，点状的彩灯效果和庭院内色彩绚丽的植物，可对视力低下和色弱的人进行一些视觉上的刺激，以帮助他们恢复；在康复中心的内庭有一条卵石步道，其中设置了台阶和扶手，是练习走路和上下台阶的训练设施；中央景区还设有坡路、远距离园路、交叉路、横向倾斜路等，为轮椅使用者提供了丰富的训练课题。

和谐的景观

成片的树荫，开阔的草坪，规则的台地简单而又自然，很好地与建筑相互呼应。各组团景观根据功能的需要展开各自的主题，但又保持着整体的协调统一，共同营造身心康复的和谐景观。

The designers introduced No Barrier concept for communication, traffic and recovery of the patients inside the project.

CARETAKING SCENERIES

The designers introduced collective and individual communication areas. The collective communication areas are located in places where people gather and individual areas find use in many places. In addition, there are a lot of outdoor chairs for the patients to use.

FAIR SCENERIES

The vehicle roads and waling paths inside the base are independent and the elevation of the walking paths is higher than that of vehicle roads. Blind sidewalk, sound reminder and map can lead the disabled to everywhere they want to go to. Gradient of the entrance path of each building is ranging from 2% to 5% grade.

RECOVERY FACILITIES

The facilities include those for audition training, vision training and walking trainings.

HARMONIOUS SCENERIES

Tree shadow, grassland and podiums are compatible with the buildings. Each complex serves their users and they all are in a united manner.

工程档案

工程名称
上海残疾人综合设施暨特奥竞赛训练基地景观设计

工程地点
上海市松江区

设计单位
上海现代建筑装饰环境设计研究院有限公司

设计师
徐琏、何鉴、林源荣

项目景观总占地
26公顷

单位平米造价
120元

设计完成时间
2007年7月

竣工时间
2007年8月

1. 东行主环岛
2. 康复中心庭院
3. 东面轴线
4. 南面主轴绿化
5. 培训中心庭院
6. 台地
7. 宅前绿化

5

6

4

7

Overall planning and design of Expo grass land, Shanghai

上海世博园绿地总体规划设计

秉承生态与人文的设计理念，本方案用“三明治”式的带状绿地结构，在一个工业用地上，恢复了一个生态的绿色基底，在此基底上建立一个便捷而体验丰富的交通网络。这个网络联结了一系列文化主题结点，讲述上海的过去、现在和未来。本方案综合解决了滨江洪涝、生态重建、人流疏散、遗产利用、文化展示等问题，使之成为一块滨江绿地、世博绿地和城市绿地，满足会展期间和会展之后的多种功能需求。

设计面临的问题

滨江洪涝：如何利用场地的滨江特点、利用雨洪和海潮作用下的间隙性水位变化，形成既安全又富于景观特色和满足使用功能的绿地？

生态重建：如何将一块工业用地恢复为生态安全、环境良好的再生绿地，为区域和场地提供生态服务功能？

人流疏散：展时和展后的人流量变化很大，如何在会展期间满足大量人流的疏散和滞留，同时又能方便而经济地转为可持续的日常游憩绿地？

遗产利用：如何保留和再利用场地中的工业遗产，使之成为绿地的有机组成和特色景观，并能满足新的使用功能？

文化展示：如何根据场地特色、配合世博主题，形成绿地的文化特色，体现过去——现在——未来？

设计应对的策略

“双堤夹镜”结构的绿色基底

这种结构很好地解决了防洪和场地使用问题，同时也会较少的动用土方。场地防洪共分为三个等级：6.7m千年一遇防洪堤克服了以往堤坝形式单一、刻板僵硬不能给人以美感的形式，堤岸后移，选址既保证打开城市与江面之间的视线联系，同时又不切断中心绿地与展馆区之间的关系；5.4m百年一遇的防洪堤与4.5m二十年一遇的防洪堤方案中均采用灵活的堤坝形式，塑造了场地丰富的空间。整个基底体现了自然与人文、生态与艺术，安静与热闹的递变与调和。

建立便捷的交通和丰富的体验路网

与绿地基地的自然和人文属性相对比，在绿地基底上构建四种道路，构成交通和体验网络。南北向的直线林荫步道，连接着展区、绿地直到江边，方便快捷。会展期间，用造价低廉而富有江南特色的

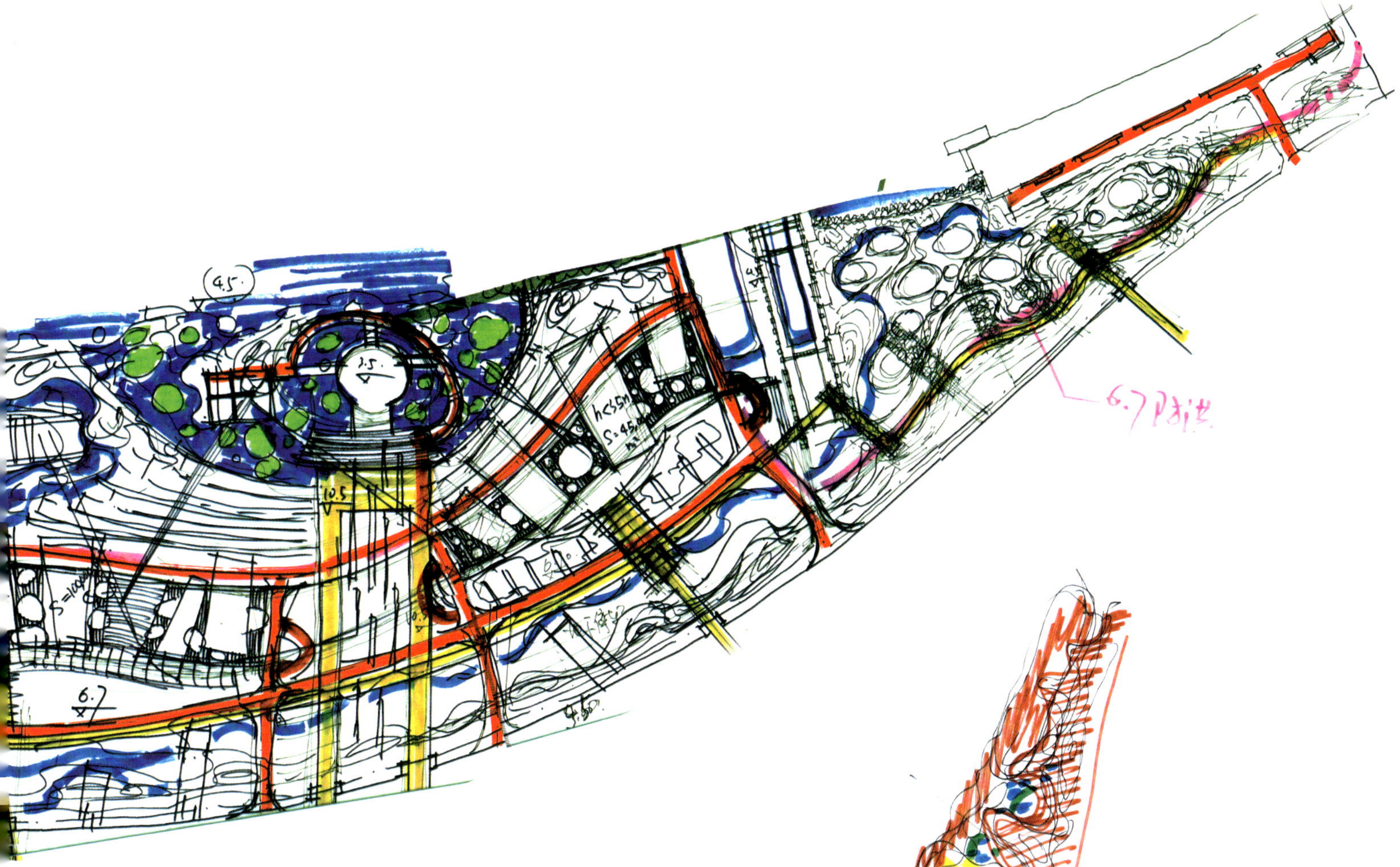

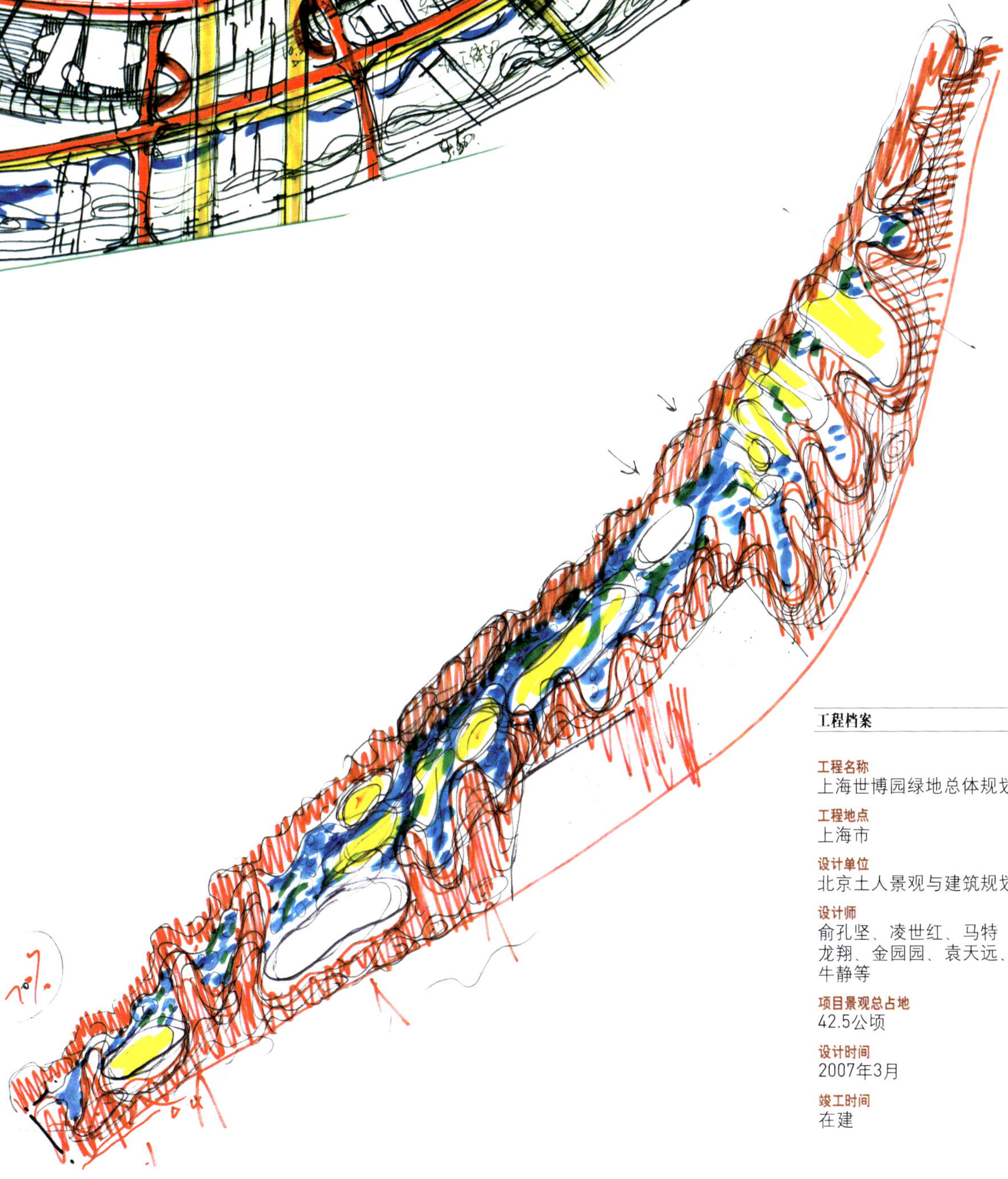

工程档案

工程名称
上海世博园绿地总体规划设计

工程地点
上海市

设计单位
北京土人景观与建筑规划设计研究院

设计师
俞孔坚、凌世红、马特（德国）、刘向军、龙翔、金园园、袁天远、蒋伟荣、冷刚、牛静等

项目景观总占地
42.5公顷

设计时间
2007年3月

竣工时间
在建

驳岸区位分析图

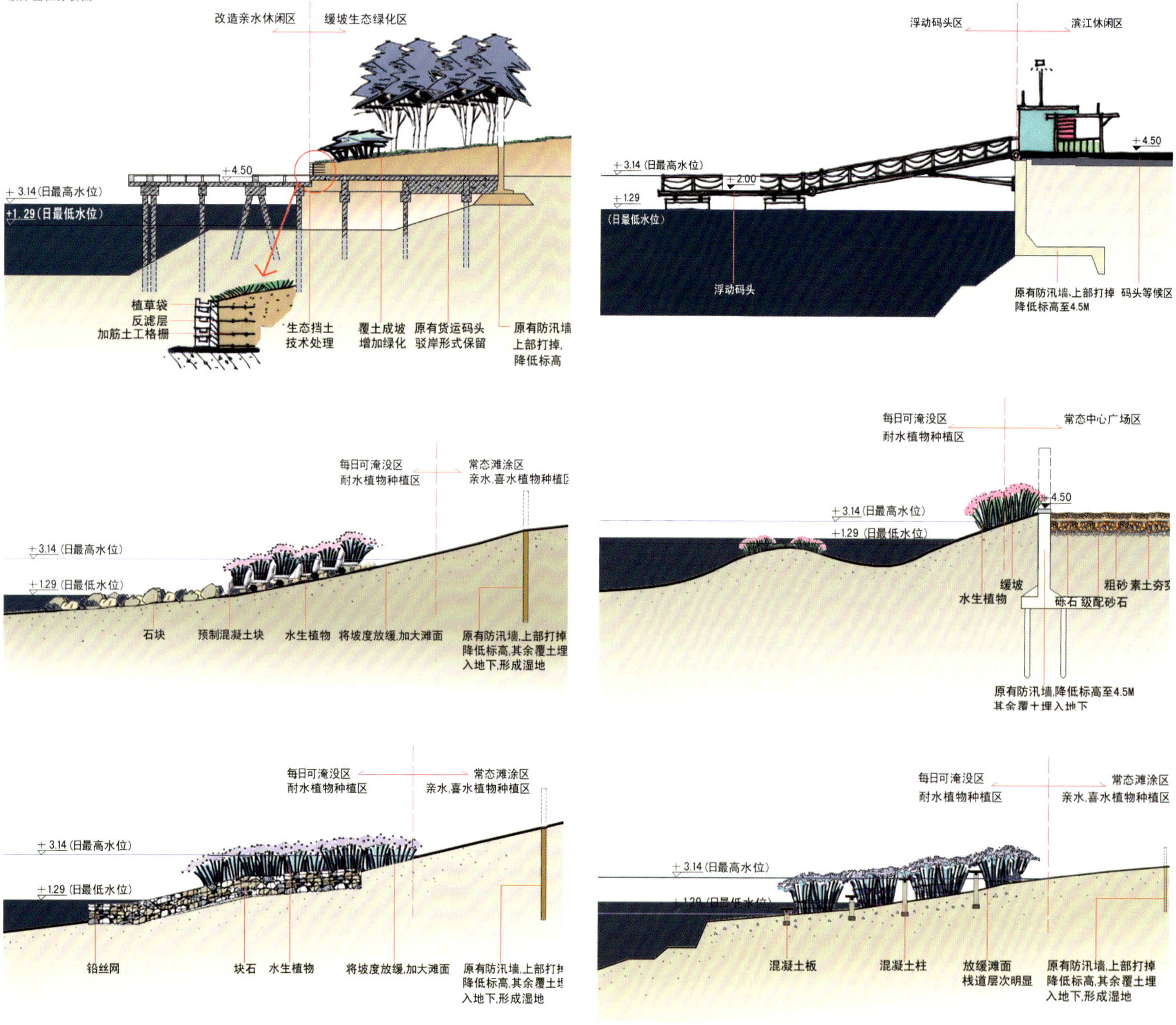

会时总平面图

透视图

船坞茶室透视图

泡状湿地分析图

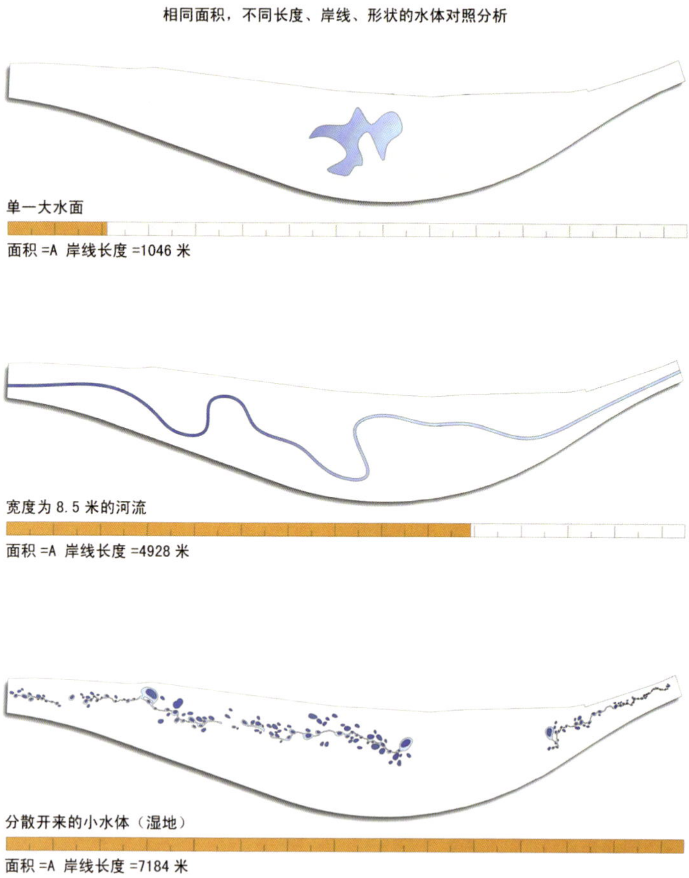

竹材铺成10米宽步道，以利大量人流疏散，展后两侧恢复绿地，竹材粉碎回土。

多个文化主题结点，展示上海城市的过去——现在——未来

绿地中分布了多个文化主题结点，包括休闲吧、空中花园、龙门吊休闲广场、世博庆典广场四个功能区、现代庭院艺术带中的15个主题艺术园。其中前三个结点都是利用现有工业遗产和建筑物改造而成，是过去城市记忆与现在都市时尚生活的有机结合。世博庆典广场则是时代精神的体现，是上海蓬勃发展和新时代海派文化的体现；而主题艺术园则是新科技和新的环境伦理对未来城市生活的预示和引导。

滨江码头与防汛堤改造利用

上海世博会作为中国向世界展示的窗口，处处都应体现对环保、生态问题的重视，从而进一步强调"城市，让生活更美好"这一主题思想。

生态与人文理念下的技术

乡土物种应用及生物多样性的保护意识及措施；材料的节约和循环利用；雨水收集、循环及处理系统；利用生物净化原理，通过岛屿结构扩大边界面积，提高生物净化效率；利用太阳能和风能来进行水循环及场地内的照明。

会期与会后的弹性设计

在设计之初就已考虑了可拆卸性和材料的可回收性，这种弹性设计不仅可以节约成本，而且也为场地的可持续发展提供了良好的基础。

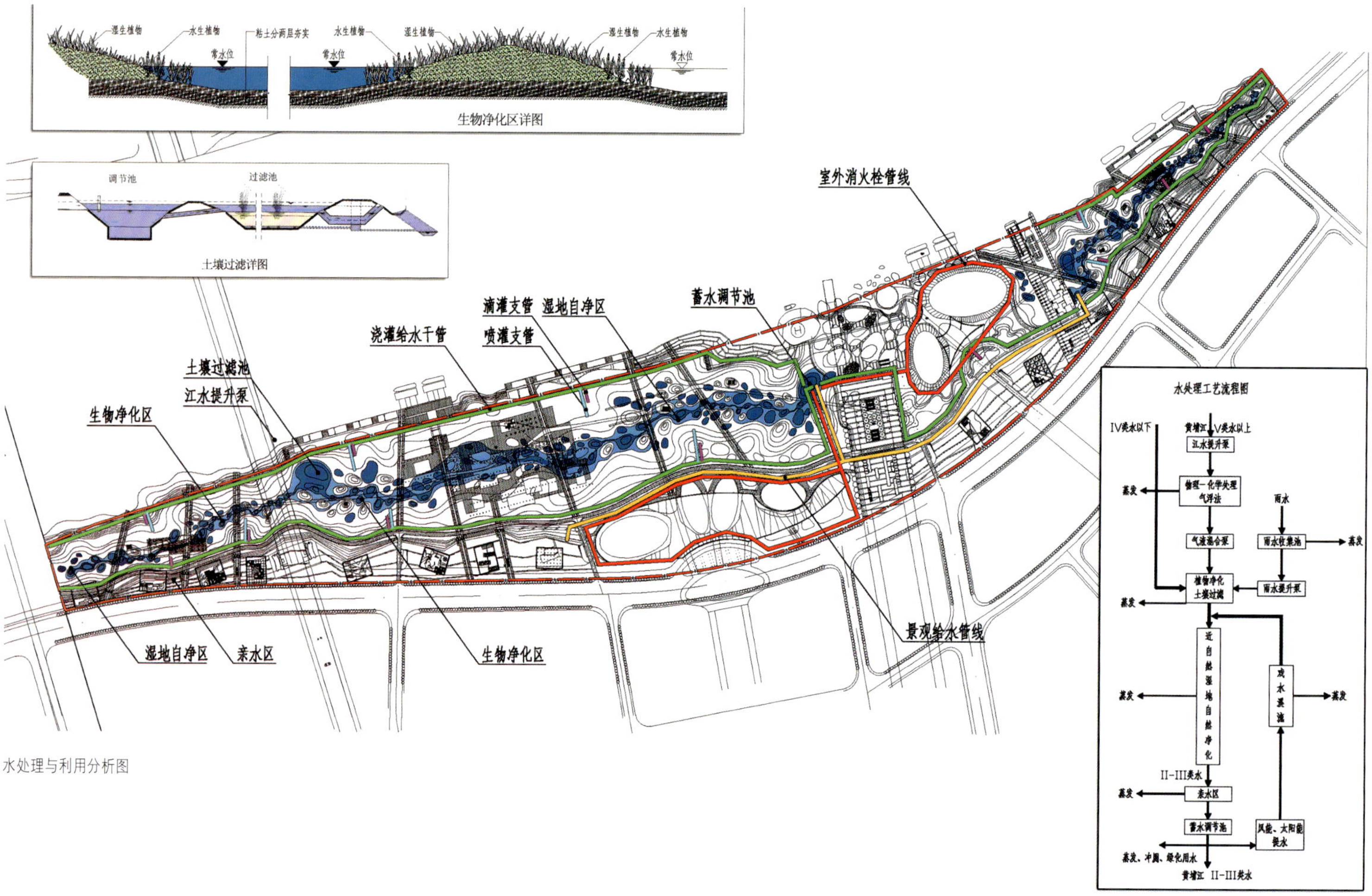

水处理与利用分析图

透视图

龙门吊透视图

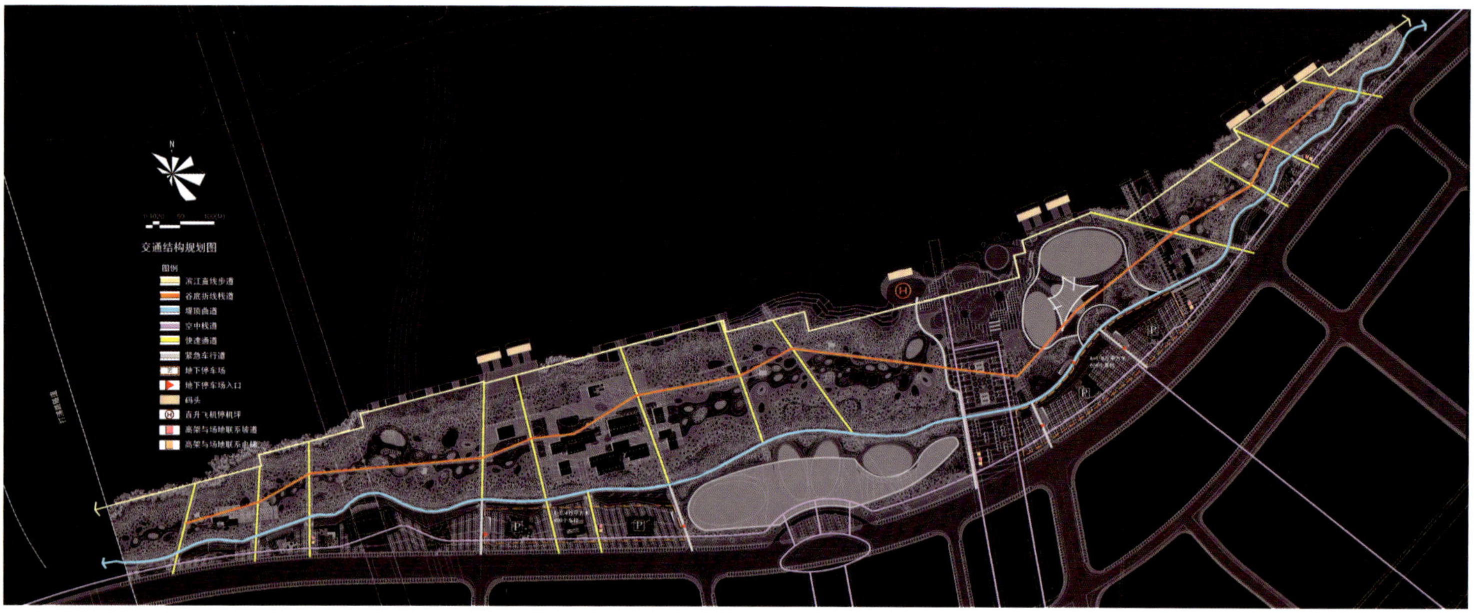

交通结构分析图

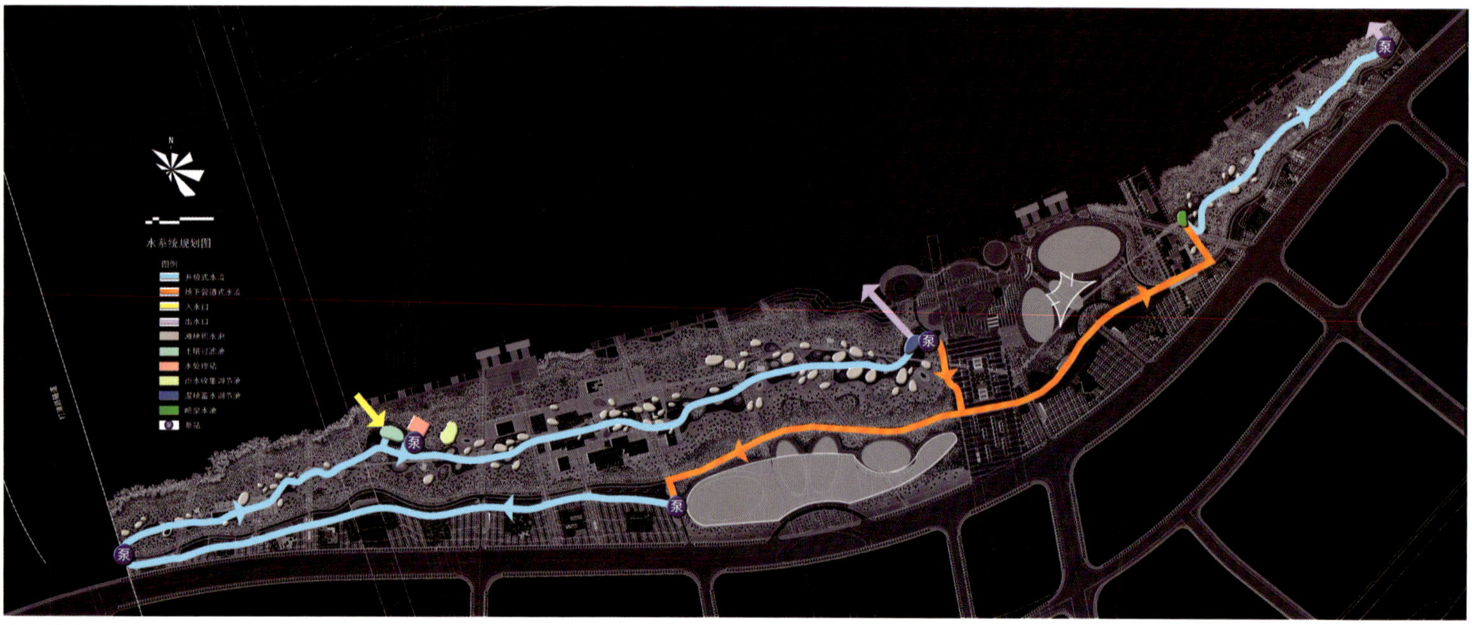

水系统规划图

湿地剖透图

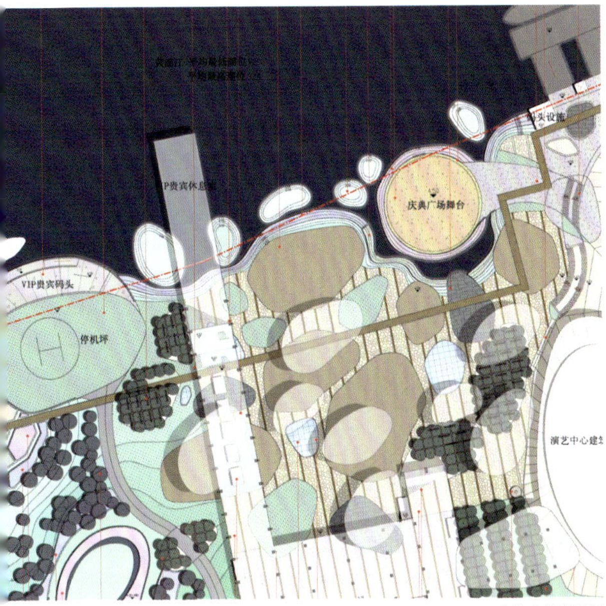

庆典广场平面图

The project adopts a sandwich like belt structure in restoring an eco base on a land used as workshops. And a convenient traffic network is built on this green base. This network connects a series of culture points on Shanghai in all the time. The project manages to solve the problems of flood, reconstruction, people dispersing, heritage reuse and culture displaying.

DESIGN METHODS

Green base

The structure has perfectly solved the problems of flood controlling and land use, while digging land as less as possible. The entire base highlights the transition and conciliation of nature and human, eco and arts, and quietness and stirring.

A convenient and enriched experiencing traffic network

4 kinds of roads are created to serve all people during the Expo event and the grass land can be restored immediately after the Expo.

Multi themes to display Shanghai in all times

The grass land houses up to 15 theme parks and 3 of them are based on existing heritages and buildings, a dynamic combination of the past city and modern fashions.

Riverside pier and transformation of the flood bank

Shanghai Expo emphasizes the problems of environmental protection and ecology, under the theme of Better City, Better Life.

Eco and humanity-based technologies

They are highlighted by various methods: conservative and recycling use of the materials; rain collecting, recycling and treating system; improved biological purification efficiency and use of solar and wind power.

Flexible design

The materials are easy to dispatch and to be recycled. The method can conserve the cost and provide a sustainable development base for the ground as well.

广场透视图

Landscape design of east bank of the Canal, Hangzhou

杭州运河天地一期景观设计

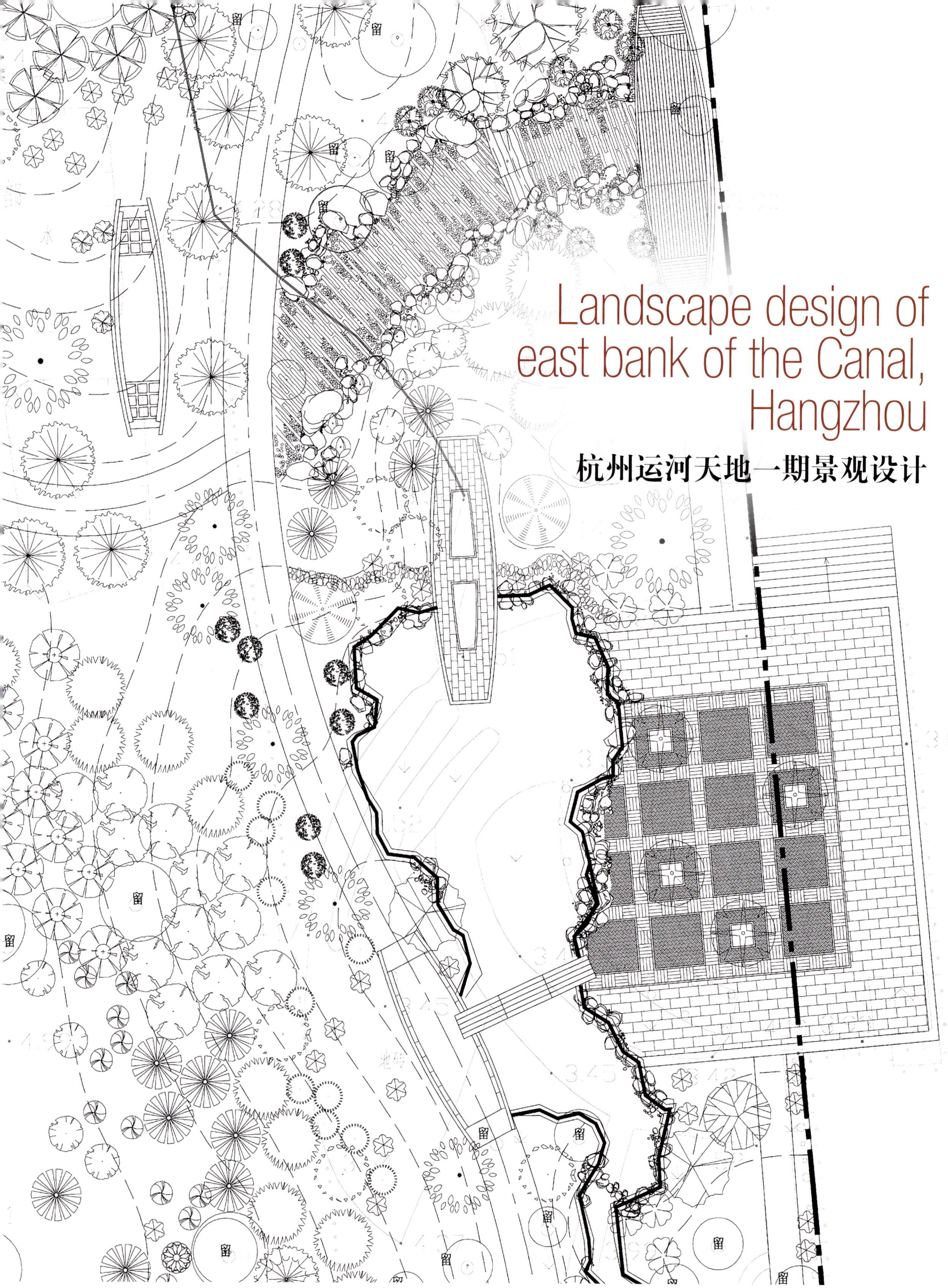

项目概况

项目位于京杭大运河西岸，一期建设用地面积约6.83公顷，规划设计围绕原址旧厂房的保留、历史文保点及休闲公园的建设展开，将旧有保留厂房改造成创意产业园，既为市民提供休憩场所，又作为运河景观的一个重要组成部分。

设计原则

本项目意在诠释城市滨水区域景观设计的新概念，用现代化的景观设计手法，表达运河景观的历史、现在和未来的意义，营造良好的人与水亲近的环境氛围。完善运河功能，使其不仅具备城市运输和排洪的功能，而且作为凸现城市骨骼脉络的水系，提升运河的旅游文化景观功能。

现状情况

地块西侧为原化工厂厂区，沿规划小河路有十几幢原有厂房建筑保留了下来，其余建筑均已拆除；地块东侧为未改造的运河绿化带和运河；文物保护建筑——高家花园位于地块南侧。现状用地范围内有许多大树，主要树种为水杉树、香樟、梧桐树以及广玉兰。

设计构思

整个项目设计包含两个主体，一是创意产业基地的景观设计，包括保留建筑的外立面、屋顶景观改造；二是沿运河的市民休闲公园。

设计依据现状，也分为两大区域：

文化创意区

该区域范围为园区主道路以西的原化工厂厂房区域，主要以改造的旧厂房建筑结合厂房周围的环境进行设计，同时增设景观小品形成室内外交融的景观空间。保留的旧厂房主要作为创意产业区的办公空间和商业休闲空间。

休闲公园区

该区设计上考虑作为文化创意区的景观大背景，不仅为基地提供良好的创作环境和展示空间，同时还为周边居民提供休闲游憩的公共绿地。公园设计为开放式公

园，步行入口设计结合原化工厂遗址和保留的植物配以景观小品，使原有厂区的工业氛围得以延续。在主要园路以西结合文物保护建筑——高家花园，设计成为以传统园林景观符号为特征的现代中式公园，在高家花园北侧设置内部循环水系，形成公园的中心水面，将沿公园主道路布置的林荫漫步道、休息平台、望鹊台、林下小憩等几处各具特色的景点串联起来，结合水系尽端设置叠水以及亲水平台。同时在休闲公园的设计上兼顾和对面创意园区的相互借景的景观设计效果。新旧对比，软质景观和硬质景观互为衬托，绿化配置也尽量结合园区保留树木，以水杉树、香樟、梧桐树以及广玉兰为主。

运河天地作为城市公共开放空间，其景观照明主要分建筑照明、硬质照明和绿化照明三部分。照明形式有泛光夜景照明、建筑物轮廓照明、内透光照明和特种夜景照明等，展现夜间运河天地的景观效果。

1. 局部保留工业建筑改造效果图
2. 局部保留工业建筑改造效果图
3. 运河天地彩色总平面图
4.5.6. 局部保留工业建筑改造后实景照片

下一跨页
1. 保留工业建筑改造后实景照片
2.3. 保留工业建筑改造后实景照片
4. 服务中心实景照片
5. 结合保留水池的休息平台
6. 保留旧厂房改造后实景照片
7.8. 休闲公园水景实景照片

1

2

3

4

5

6

7

8

工程档案

工程名称
杭州运河天地一期景观设计

工程地点
浙江省杭州市

设计单位
中国美术学院风景建筑设计研究院

设计师
武荷茌、沈赟、裘常恩

项目景观总占地
6.83公顷

单位平米造价
370元

设计完成时间
2007年3月

竣工时间
2007年10月

1. 主入口钢架构筑实景照片
2. 景观栈桥实景照片
3. 保留厂房改造后实景照片

1

2

PROJECT BRIEFING

The project is located at the west bank of the Beijing-Hangzhou canal, covering an area of 6.83 hectares. The project is expected to maintain the existing workshop, historical sites and build a sports park.

DESIGN PRINCIPLE

The project is expected to reinterpret the concept of waterside scenery design and modern approaches will be used to explain the riverside landscapes. The canal needs to be upgraded for traffic, flood and becoming new riverside sceneries as well.

CURRENT SITUATION

There are some workshops at the west side of the land and on the east side are unimproved greening belt and canal. Gao's Garden, a protective building, is located in the south end of the land. There are a lot of trees on the land.

DESIGN CONCEPT

The entire project includes two groups, one is the scenery design of the Innovation Industry base, referring to maintaining the outer facade of the buildings and improvement of the house roof. The other is a sports park alongside the canal.

Culture innovation area

The old workshops and their environment need entire redesign, supported by scenery ornaments. The workshops can be used as offices and shopping centers.

Sports park area

The area is expected to be a scenery background of the culture innovation area, while serving residents around it. The design plans to sustain the workshop environment here.

The Gao's Garden will be built into a modern Chinese style garden. There will be a central water park in the north end of the garden, and series ornaments are connected alongside the central forest path.

The lighting systems are designed for buildings, hardware and greening respectively.

3

Landscape design of Tongjun Plaza, Tonglu, Hangzhou

杭州桐庐桐君广场景观设计

设计理念

桐庐——一个拥有深厚历史文化渊源的城市，桐君广场——一个具有崭新商业策略的城市公共广场。如何将文化传承、自然景观与现代商业策略融会贯通，成为本设计的关键。广场的景观设计打破常规，以一组组半围合式的绿色圆环为线索，把交通、人文、历史、自然等碎片有效地组织起来，并进行重新演绎，使得整体的景观环境呈现与众不同的特点与氛围。同时，把景观作为文化的载体，在设计上将文化简约的运用，从材质到小品都展现了文化，大文化的概念和精神在小细节的景观设计中得以体现。

设计方式

在充分尊重现状的基础上，挖掘历史文脉

历史上的小东门又称启明门，地处富春江和分水江的交汇处，在原有东城门的旧址上拟运用现代的理念反映当时的历史风貌和老渡口的台阶码头，并结合西侧的八角亭建设沿江的传统风貌，为此地区注入更多的人文历史内涵。

积极研究基地独有的自然元素和材料，并使之获得再生

保留、再利用基地内的景观元素和材料，使得他们能发挥新的使用与审美功能

1. 停车区
2. 民族文化街区
3. 民族文化街小巷
4. 庭院

1

2

是我们本次设计的基本思想之一。利用乡土材料和本土植物，使新的设计对环境的负面影响做到最少，利于整体环境的改善，并融入装置艺术等多元现代文化气息，比如：铁锚、浮标等均被巧妙设计成为桐君广场景观元素的一部分。

不拘泥于传统形式语言，而是采用更为简约、内敛、鲜活的设计手法

通过适当的材料、合理的工艺以及独特的形式，延续主体中心广场（海螺形广场）的构成方式，同时加入交织的设计手段，错落有致、旋转的波纹有机地和整体环境相结合，把基地塑造成高品质的景观环境。

拓展新的城市功能点，创造新的城市景观亮点

由于位于城市一隅，与城市的关系日益疏远，设计拟将增加新的城市功能和设施，将街道和广场延伸到码头文化区（原小东门位置），并与民俗文化区巧妙的结合，从而创造新的、有品质的滨水景观区，通过自然山水、建筑、标志物、桥梁、台阶、不同层面的平台、绿化与小品等的交融和渗透，塑造立体的、多层面、多形态的空间体系。

3

4

DESIGN CONCEPT

Tonglu is a city with rich historical and cultural resources and Tongjun Plaza is one with fresh new commercial strategy. The key of the design is how to integrate culture heritage, natural scenery and modern commercial concept. The design of plaza gives up practical norm, rearranging the factors of traffic, human, history and nature with a group of semi-circular greening belts. The result is that the entire scenery environment is different from others. Meanwhile, sceneries are used to ship culture that is embodied in terms of the materials and ornaments.

DESIGN APPROACH

To explore the culture relics with full respect of the current situation

The Xiaodongmen used to be located at the cross of the Gongchun River and the Fenshui River. Modern concept is planned to be used on the site to reflect the historical touches and steps of the old ferry. The constructions alongside the rivers are set to insert more human and historical contents.

To rejuvenate the peculiar factors and materials

The approach is one of the basic ideas of this design. The local materials and plants are used to impose negative effects as little as possible and therefore anchor and buoy have become part of the scenery factors of the plaza.

To adopt design approaches which are pact, restraining and energetic

Materials, techniques and forms are used to sustain the composition of the central plaza (trumpet shell shape), while adding interwoven design approaches. Revolving ripples are integrated with the entire environment, making the base a high-caliber place.

To create new scenery spot

Because the spot is moving away from the central city, the designers plan to add new functions and facilities: the roads and plaza are extended to the pier culture area, combined with the folk-custom area. Various factors are used to create three-dimensional, multi-layer and multi-shape space series.

工程档案

工程名称
杭州桐庐桐君广场景观设计

工程地点
浙江省杭州市

设计单位
中国美术学院风景建筑设计研究院
现代设计研究所

设计师
王昀、赵阳

项目景观总占地
3公顷

单位平米造价
400元

设计完成时间
2007年5月

竣工时间
2007年11月

1.3. 中心广场
2. 主入口步行街
4. 集团办公楼

2

DESIGN CHARACTER

Plants and quality

Point planted plants are used to arrange the traffic routes of the entire plaza, making the single green land and circular island energetic. The flagstone and wooden platform make people feel comfortable while shopping or walking. Arbor tree, chair and floor are reminding people where they are.

Sloping field and environment

Sloping fields in the folk-custom plaza help make the design enriched and enjoyable, thanks to their varied length, direction and width. Sloping fields of different grade and length lead to the round plaza from river sides. And the plants help make the sceneries solid and penetrative.

Space and life

The plazas and wooden platforms are prepared for entertainment activities and the private rest areas and peculiar buildings are reminding people the meaning of the street. The design also makes the residents feel that they are part of the sceneries.

Pier and ornament

The old vehicle-related facilities are reconstructed: buoys are used by children; tree groups make the space among each area flexible; the broken walls are not only symbols but also transitions of the space. Wide greening lands and down steps are telling stories happened at the side of a pier.

3

4

1. 艺术馆透视图
2. 会客厅庭院透视图

Landscape design of Lantianyiyuan residential community, Shaoxing city

绍兴蓝天怡园景观设计

项目概况

此项目地块北面临水，设计范围内现状用地主要为林地和空地，另有少量民居及饲养场建筑。基地内有一条南北向的水系和一条东西向的水系，与北面河流贯通，在地块西侧有一个小型池塘。设计范围内植被较茂密，有大量的灌木及多年生高大乔木。规划中要大力保护古树，努力营造一个富有天然情趣的健康生态小区。地块原始生态良好，有天然的鱼类及鸟类。地块周边环境较好，只是东侧的华安公路噪声较大。

布局结构

本案规划布局结构可概括为“三区二园”。三区：由东侧沿街公寓、中间排屋及西侧公共建筑三个区块共同组成一个完整的社区环境。二园：中部的排屋区、西侧的公共建筑区形成的两个相对独立的江南园林格局。

交通组织

公寓区人流通过东侧华安公路直接进入院内，人行系统与排屋区和公建区互不干扰；排屋区及公共建筑区内设置必要的消防通道以满足消防要求，力求建立比较完整的内部人行系统，并通过连廊、平桥、拱桥、廊桥、汀步将所有建筑有机结合到一起。

景观设计

中国江南古典园林注重“景无情不发，情无景不生”，怡园在规划和建筑设计阶段就遵循了古典园林的布置手法。亭、台、

1

2

楼、阁、桥、轩、廊、榭，错落有致，最大限度的给园林留予空间，在怡园的造景中，努力营造静谧的空间，让置身园中的人能在静中求动，寂中求欢。

怡园在景观设计上注重布置与处理好空间，即在“空”的地方，也要达到“虚中有实，实中有虚”，如用水面衬托庭院，延伸空间，以虚代实；借用粉墙映衬树影，产生虚实相生的景观。园中大多数庭院面积是有限的，因此，布置时常用借景和空间渗透来扩大视野。同时汲取山水画中建筑与树木交叠的方法，园中树木排齐，以屋宇间之，屋后再做树木，层次丰富。总之，在怡园的园林景观设计中，力求用现代化的设备、崭新的元素以及传统的手法再现江南古典园林的意境。园内植物叶、花、果的形状和色彩随季节而变化，具有较高的观赏性。

1. 博物馆东立面图
2. 博物馆南立面图
3. 风水分析图
4. 沿街大门立面图
5. 公寓楼南立面图
6. 大门南立面图
7. 公寓楼东立面图
8. 主要庭院透视图

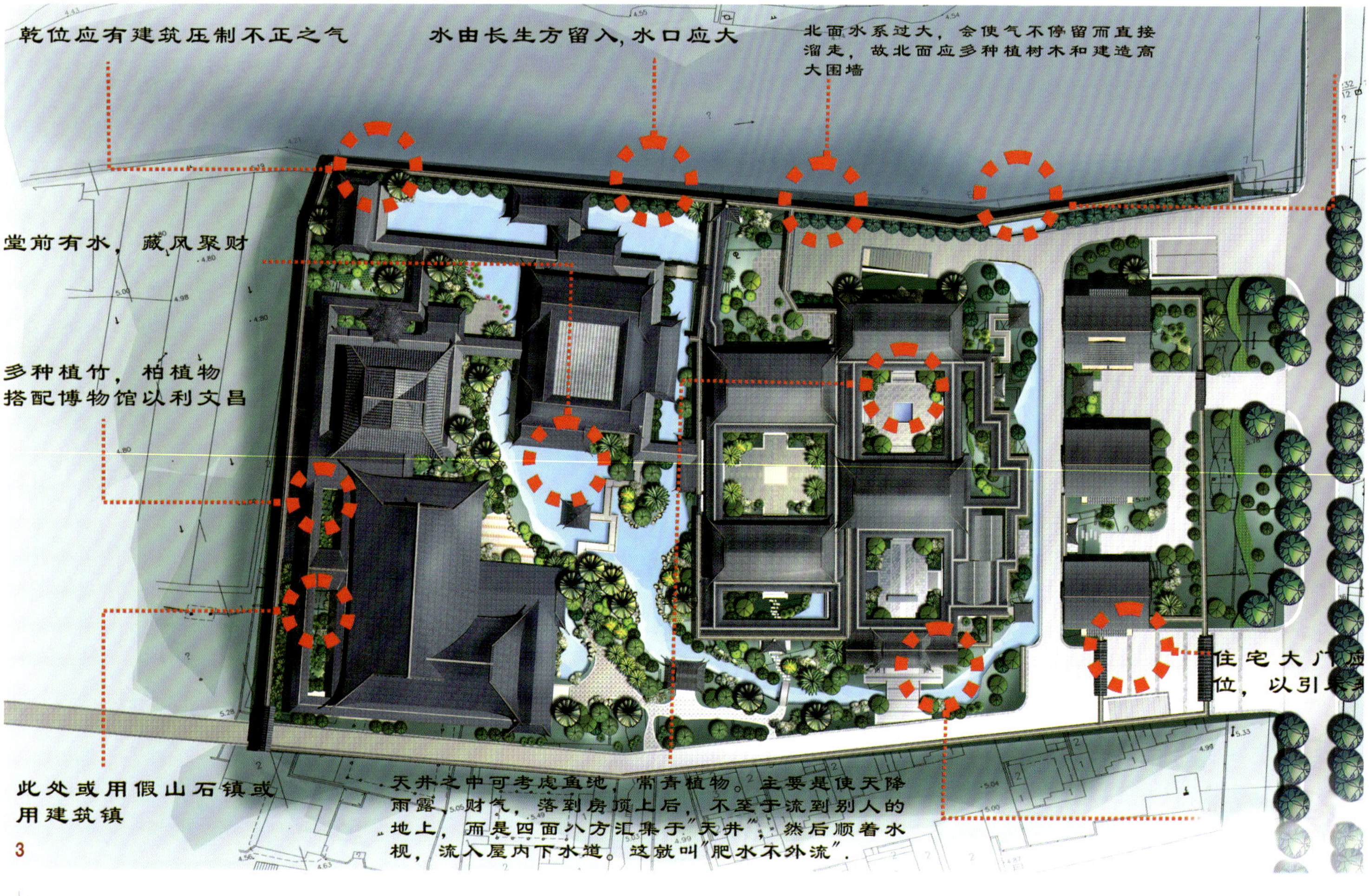

4

5

6

7

8

怡园

1

2

BRIEF SITUATION

The land, facing water in its northern side, accommodates a forest, some houses and a farm house. Two rivers are flowing towards another one in the north end and there is small pond in the west end. The thick plants include frutex and arbors. Fishes and birds are found inside the area. The land has a sound environment except noise from a highway in its east end. The design is expected to protect the ancient trees while creating an eco community.

STRUCTURE LAYOUT

The plan consists of a residential area and two independent southern Chinese style parks.

TRAFFIC ARRANGEMENT

Residents of the apartment go to inside the yards through the east entrance and the sidewalks are away from those of the other two. There are firefighting accesses in the other two sections. There is an enclosed sidewalk system, connected by all kinds of tactics.

SCENERY DESIGN

The design fully respects the traditional Chinese style gardening tactics. All buildings are well arranged, providing as much space as possible for the gardening.

The scenery design pays great attention to the relationship of space, making the limited space wider.

The entire design adopts modern equipment, fresh new elements and traditional tactics. The shapes of colors of plants change with that of the 4 seasons.

1. 主出入口透视图
2. 南立面图

工程档案

工程名称
绍兴蓝天怡园景观设计

工程地点
浙江省绍兴市

设计单位
杭州现代机构

设计师
郭宏强、周彤、干丽丽、郑碧峰、姚玉云、蒋剑静

项目景观总占地
5.6公顷

单位平米造价
300元

设计完成时间
2006年11月

竣工时间
2007年10月

Design and plan of Taiyangxingcheng, Changsha

长沙太阳星城规划设计

项目概况

项目位于长沙市开福区青竹湖外商城中的生态居住园区内，西、北、东三面为山，南面是青竹湖大道。地块原生植被保护良好，北面有出水量达每小时300立方米的泉眼，规划时保留大片生态绿地，以营造小区依山傍水的生态景观；在不改变上层规划的主要道路基础上进行设计，以便与周边路网衔接；宗地为山地和坡地地形，地势起伏，为营造丰富的竖向趣味空间提供了很好的条件。

设计构思

设计总体布局以“风生水起”的自然轮回来酝酿方案，充分体现对自然的尊重。以“水”为主线，溯流而上，将各区串联形成有机的网络，包含了文化和生态的内涵。将山景引入小区，使建筑的起伏与山势的走势和谐，建筑与山体融为一体，并创造观山看水的视线通廊，人入其中，如在画中；将现有的池塘、水系连通为活水，成为串联整个小区的绿带项链，为铸就生态水岸居住环境创造条件；将景观要素辐射到各个住宅组团中，让各家各户欣赏到最好的景观；整个小区与青山绿水融为一体，建筑仿佛是从山中自然生长出来的。

景观规划

第一重景观包括大绿洲、公共广场、生态岛、公共滨水空间等，是社区的公共交往空间；第二重景观包括组团绿地、游园，为半公共小绿洲景观，被组团居民使用；第三重景观包括院落绿地、入户花园，为每户独立享用的私密空间。

通过对建筑和景观的合理组织，打通若干条看山观水的视线通廊，拉长景观轴线，减少建筑对人的压迫感，使整个小区空间

通达舒展；由入口商业街、公建化的酒店式公寓、高低错落的广场空间引入宽阔的湖面，空间豁然开朗；在湖岸设置若干木栈道和休闲平台，增加行走湖岸的乐趣，满足人们亲水愿望。

景观分区

以天籁之域，风韵之都为景观主题，取植物、自然现象、山水、楼阁为造景要素，寓情于景，将小区分为六大主题景区：天景园、风荷园、竹溪园、清风园、静声园、太阳谷。

道路规划

主要车行道为环状系统，道路标高基本上在同一等高线上移动，最大程度地减弱对自然环境的破坏。二级路、三级路则结合地形及建筑群组自由布置，形成有机的道路网络结构。

工程档案

工程名称
长沙太阳星城规划设计

工程地点
湖南省长沙市

设计单位
爱普斯顿国际

设计师
王艳华、宋芸芸、欧阳琦、周波平、柴俊、周俊林、林越峥、曾繁强

项目景观总占地
51.1公顷

设计完成时间
2007年12月

竣工时间
2009年

PROJECT BRIEFING

The project is located in an eco living area of Kaifu dist, Changsha, neighboring hills in its west, north and east sides and the Qingzhuhu avenue in the south side. The natural developed plants are well protected and in the north end there is a spring mouth which produces 300 cubic meters of water per hour. The project is expected to maintain the large areas of grass land; make the roads accessible to the neighboring roads smoothly; and create vertical space.

DESIGN CONCEPT

The design fully respects the nature. Water is used to connect all sections; to introduce hill sceneries into the community, integrating buildings and hills; to vitalize the existing ponds and waters; to make scenery elements available for every house; and to integrate the entire community into the hills and water, making the buildings seem like being built naturally.

SCENERY PLANNING

The first layer refers to large oasis, public square, eco-island and riverside space for public communications. The second layer refers to complex grass lands, tourism garden for complex residents. And the third layer refers to private grass land and house garden.

The rearrangement of the buildings and sceneries form some visionary corridors, making the space open and wider. People are led to a wide lake by the entrance shopping street, apartment and squares. Alongside the lakeside are wooden roads and podiums.

SCENERY DIVISION

The elements of vegetation, hill, water and kiosk help cut the community into 6 sections.

ROAD PLANNING

The leading traffic roads are enclosed and of the same contour level, impacting the nature as less as possible. The sub-roads are compatible with the landform and building complex, making the network organic.

Landscape planning and design of relocation project of Ditan Hospital, Beijing

北京地坛医院迁建工程景观规划设计

北京地坛医院是我国规模最大的传染病专科医院之一。为完善北京市突发公共卫生事件医疗救治体系，北京市委、市政府决定迁建北京地坛医院，并将新建的北京地坛医院定位于“亚洲一流的、现代的、综合性的传染病医院”。

总体布局

洁污分区

由于北京主导风向是西北风，项目用地由北至南呈狭长形，所以，建筑规划设计将非医疗功能区布置在用地的北侧，污染源较多的医疗功能区布置在城市常年下风向——用地的南侧。根据医疗流程和传染病学的防护要求依次为急诊医技部、一般传染住院部、呼吸科住院部。

功能分区

针对传染病医院的特殊性，设计需从传染源、宿舍、环境的三角形关系中切断传染链，控制传染源。依据北京市常年风向及建筑总体布局，明确景观环境的各功能区及洁污分区与流线。由北至南依次分为洁净区、缓冲区、污染区。其中北侧的洁净区主要为医护人员服务，景观功能区包括：健身花园、农艺花园、休闲花园、雕塑花园。南侧的污染区主要为住院的病人服务，景观功能区包括：天使花园，疗养花园，感官花园。中间的缓冲区主要是病人及其家属出入的地方，是一个敏感的临界区，需要特殊处理，景观功能区包括：主入口广场、林荫广场、下沉花园。所有景观功能区之间均有独立的出入口和明确的缓冲隔离措施，如树阵、绿篱、围栏、暗藏式自动门。做到洁污分区、洁污分流明确，互不穿插交叉，以利于防止与控制院内交叉感染。

交通流线

医院出入口和交通组织基本布置要本着划分洁污分流的原则，对可能成污染源的流线严格控制，尽量短捷，保证院区环境安全。其他各种主要流线要保证人流物流顺畅便捷。根据医院功能分区，医院设置了六个出入口。即主出入口（病人出入口）、次出入口（医务工作人员出入口）、后勤物资出入口、污物出口及呼吸科应急出入口。

洁污流线

急诊医技部、各住院部门诊、急诊患者人流、住院患者、探视人员主要出入口布置在建筑东侧，通过入口广场与主入口连接，尽量减少病人院区内的活动区域。医院各部门医务人员出入口均布置在建筑西侧。外购物品的接纳验收，配送物流发送，污物的存放运输安排在地下解决，减少各种物流与地面人、车流的交叉。物品供应车流定期、定时、定路线作业。废弃物车流夜间作业，由专用密封车辆完成，降低了污物对整个院区环境的影响。东侧道路中部连接病人出入口，安排各类病人及探视人员流线。西侧道路连接医务工作人员出入口、后勤物资出入口，安排医务人员、清洁物品流线。

紧急预案——四级警界防护隔离体系

在院区的30m隔离带外围均设定隐藏在绿篱及树林中的围栏，出入口处均设有自动门。建筑之间的连廊与道路的交叉点

1.2. 入口广场效果图

处设计自动门，并由此做了四级警界防护隔离体系，实行分级控制。分级控制由低至高分为A级（绿色警戒）、B级（黄色警戒）、C级（橙色警戒）、D级（红色警戒）四级控制措施，实施动态管理。

A级(绿色警戒) 通过关闭连廊及控制点内的伸缩门，保护办公区域不受污染。

B级(黄色警戒) 将门诊部作为控制重点，将办公区、门诊部及住院部分隔开，减少交叉感染。

C级(橙色警戒) 严格控制住院部，保护其他区域不受污染。

D级(红色警戒) 通过伸缩门等强制措施，严格控制呼吸科住院部，并将呼吸门诊临时迁至住院部，控制疫情恶化。

通过以上四级应急措施能够完全避免院内的交叉感染。

种植规划设计

由于地坛医院的特殊医疗任务，绿化隔离带拓宽为30m，这从基础功能上满足了隔离空气的要求，同时为营造出一条自然林带提供了有利条件。其中位于行政楼的西北角的健身花园，种植遮荫能力较强的乔木：槐树、栾树、泡桐，灌木选用无刺的较为安全的植物：紫丁香、大叶黄杨、碧桃、金银木，搭配上麦冬、萱草等地被植物，让人在运动的时候同时心情舒畅。医院中间的缓冲区，主要是病人及其家属出入的地方，需要特殊处理。以种植有益胃、肾经的植物为主，如石榴、红叶小檗、贴梗海棠、马齿苋、樱花、桃等。 南侧的污染区，主要为住院的病人服务，其中天使花园与疗养花园以消化科患者为主，因此采用有益肝经的树种，如臭椿、国槐、垂柳、芍药等。感官花园以呼吸科患者为主，多采用有益肺经、心经、肾经的植物，如白玉兰、紫玉兰、银杏、柿、连翘、樱花、桃、麦冬、牡丹等带有色、香、味的植物，让病人能够通过感官直接感受大自然的美好。

景区设计

按照景观总体规划概念，整个院区从北到南分为三个区，即北边以行政楼为主体的洁净区，中间以门诊楼为主体的缓冲区，南边以病房楼为主体的污染区。其中洁净区设计了雕塑花园，健身花园，农艺花园，休闲花园四个主题区（雕塑花园中有部分下沉花园）。缓冲区设计了主入口广场，林荫广场，下沉花园三个主题区。污染区设计了疗养花园，天使花园，感官花园三个主题区。

我们在地坛医院的景观环境设计上，既满足传染病医院的所有功能要求，又努力营造出家园的环境氛围，以达到医院环境与家园环境有机和谐的融合。

1

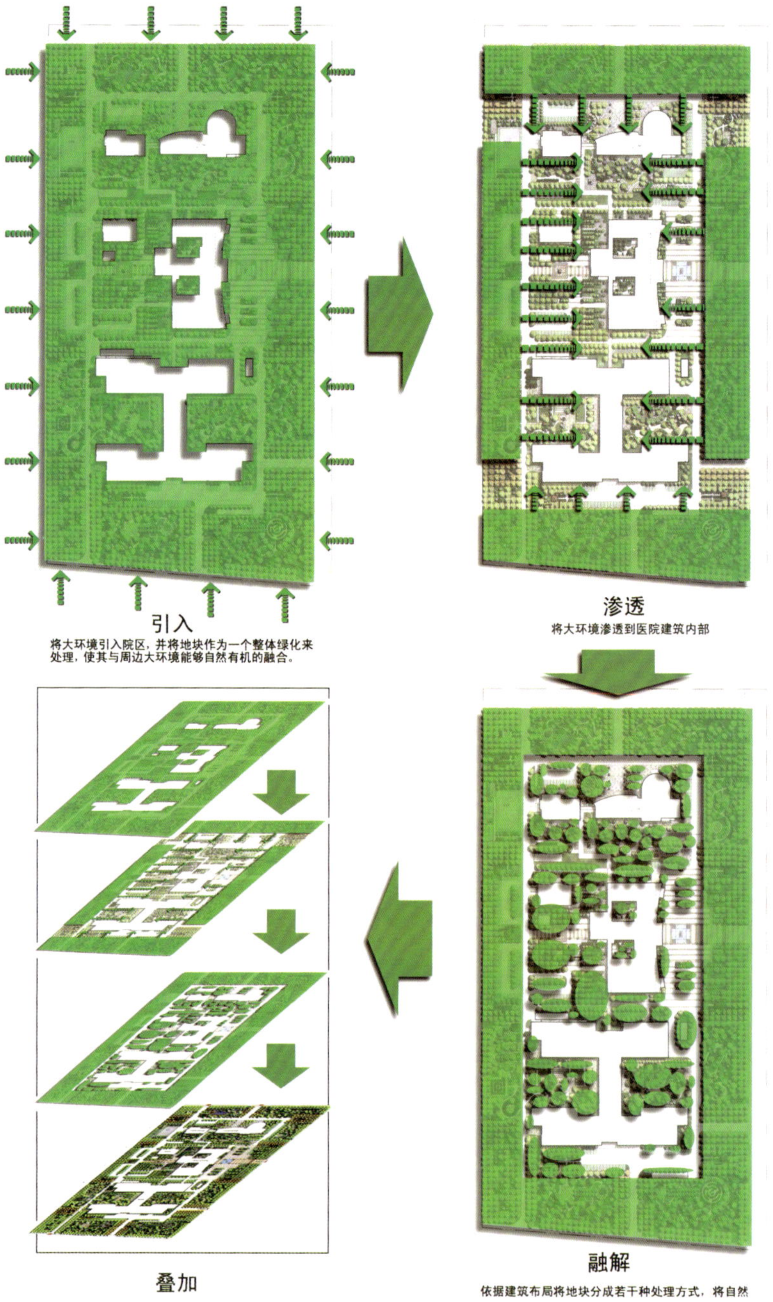

2

雕塑花园平面图

1. 总平面图
2. 空间概念分析图
3. 使用人群分析图
4. 局部剖面分析图

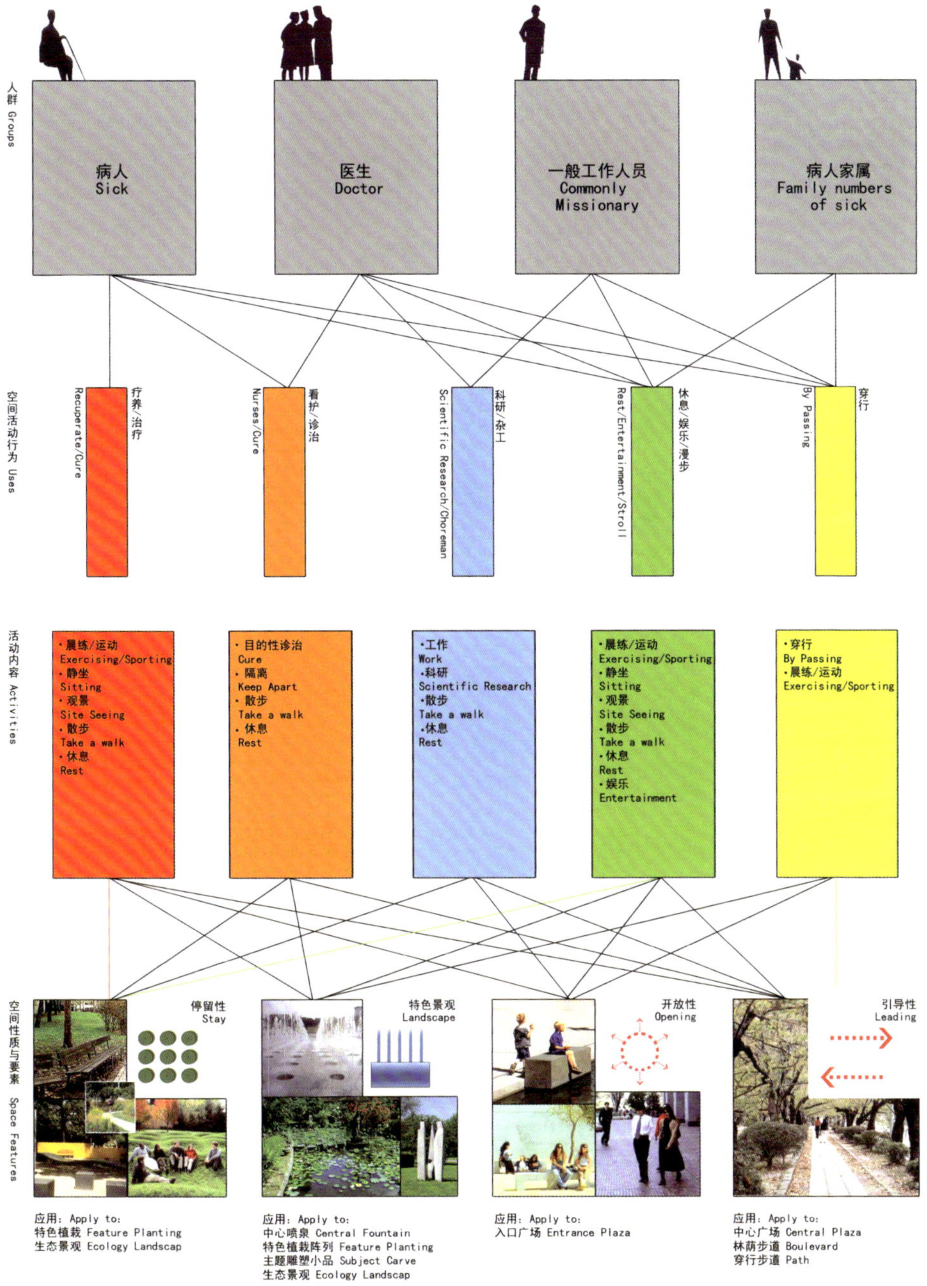

3

4

A-A 剖面图

工程档案

工程名称
北京地坛医院迁建工程景观规划设计

工程地点
北京市安外大街

设计单位
中国建筑设计研究院环艺院景观所

设计师
史丽秀、李存东、赵文斌、雷洪强、关午军、陈英夫、姜蕾

项目景观总占地
4.4公顷

设计完成时间
2007年2月

竣工时间
在建

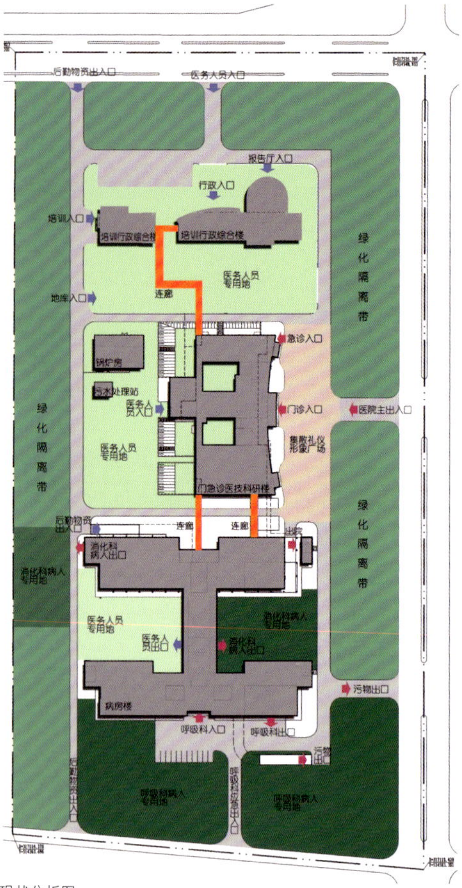

现状分析图

Beijing Ditan hospital is one of China's largest infectious institutions for 36 kinds of diseases except tuberculosis. It holds a lot of training and medicine clinical verification jobs as well. To better serve the Beijing residents, the local authorities have decided to relocate the hospital which is expected to be of "leading position in Asia, modern and comprehensive".

ENTIRE LAYOUT

Division of clean and contamination sections

As west and north end wind is common in Beijing, the plan puts the non-clinical areas in the northern part while those with heavier contamination sources in the southern part, namely they are clinic, emergency treatment, normal infectious ward, and respiration ward.

Function sections

They are clean, transition and contamination sections from north to the south. The clean section mainly serves clinical staff. The south end contamination section mainly serves in-patients and the transitional section in the middle place are for patients and other people which require special treatment. All three sections are supported by a garden. And all the function areas are enclosed by independent accesses and distinctive measures to prevent cross infections.

Traffic assesses

There are separated accesses for patients, clinical staff, logistics, contaminated stuff and respiration emergency.

Separation of the clean and contaminated mass

Entrances for those other than clinical staff are at the end side of the buildings and entrances for clinical staff are at the west sides. And logistics and contaminated stuff use the underground routes.

Emergent measures(4-step system)

Barricades are built outside the 30 meter separation belts and automatic doors are installed at the entrances. Corridors connecting buildings

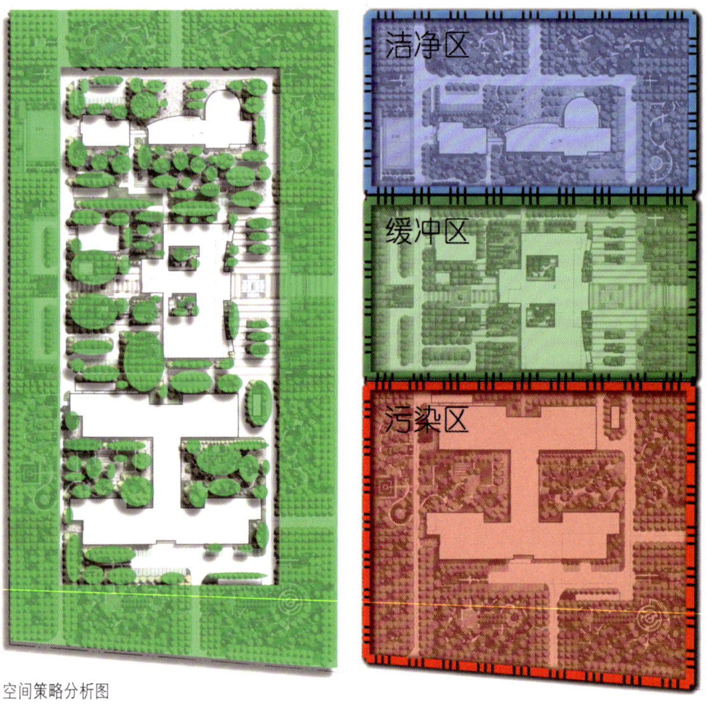

空间策略分析图

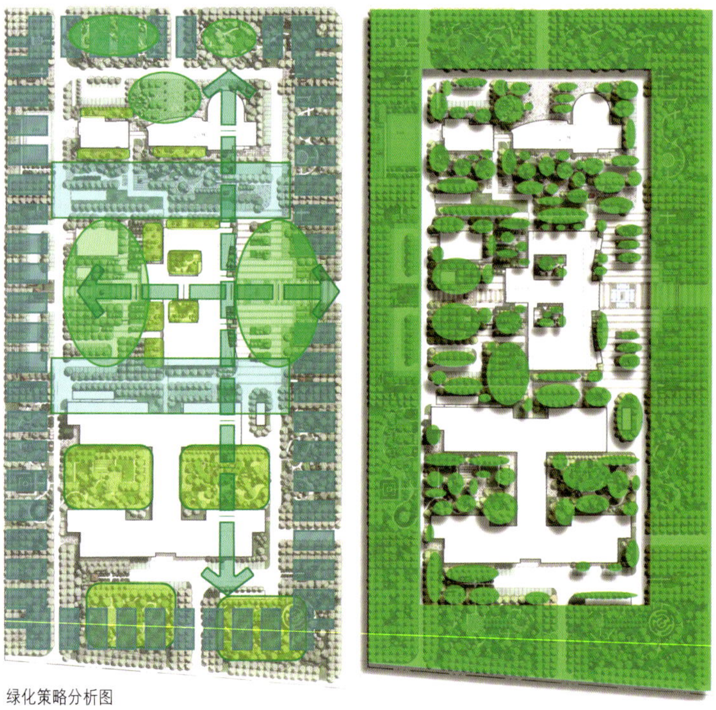

绿化策略分析图

空间策略——"化零为整"和"化整为零"，树阵林分隔洁净区和污染区

整个院区环境采用"化零为整"和"化整为零"相结合的设计手法。"化零为整"是将大环境引入院区，并将地块作为一个整体绿化来处理，使其与周边大环境能够自然有机的融合。"化整为零"是将医院环境渗透到建筑内部，并依据建筑布局将地块分成若干种处理方式，将自然环境和谐相融。在门诊楼周边东西向种植树阵林分隔洁净区和污染区。院内规划的车行路采用景观道路的处理手法，形成控制院区的空间框架，通过对景手法的处理，形成一条条景观控制轴线。

绿化策略——隔离空间、美化环境、园艺治疗、高覆盖率

绿化景观是医院的主体和骨架。方案充分尊重规划中建筑周边境的绿化防护隔离带的功能。在隔离带用地范围内有节奏、有规律的种植大量抗病毒性能强的乔木林。使建筑像是自然生长在绿色怀抱中一样。在离建筑出入口最近的隔离带内，设计满足不同使用人群的主题的庭院。各主题区根据不同使用者的需求，按照其产生的不同健康效益并按方位分别种植有利于人体经脉的植物。如在满足呼吸道病人使用的花园中多种植有利于心脉、肺脉的植物，在满足消化道病人使用的花园中多种植树有利于肝脉、脾脉的植物。这种人体保健植物群落结构，既能散发各种对人体有益的气体，调动人们情绪、增进健康，又能创造符合空间使用功能和观景需要的绿化景观。让病人不但能在绿色中放松，还能在绿色中治疗。正如南丁格尔所说的："自然能够治病，我们必须借助自然的力量"。

have automatic doors at the joint points with the roads. The 4-step separation system, from low to high, is A (green), B (yellow),C (orange) and D (red). The dynamic system is able to prevent the cross infection inside the hospital.

DESIGN OF PLANTS

The green separation belts are 30 meters wide, being able to separate air while forming a forest belt. The exercise gardens choose safety and ground plants to ensure comfortable feelings. The transitional section chooses plants good for stomach and kidney. The south end contaminated area, serving in-patients mainly, chooses more plants good for health.

DESIGN OF SCENERY

The scenery and environment designs of Ditan hospital try to integrate hospital and household living, rather than the traditional hospital measures, to provide a comfortable surrounding for patients and clinical staff.

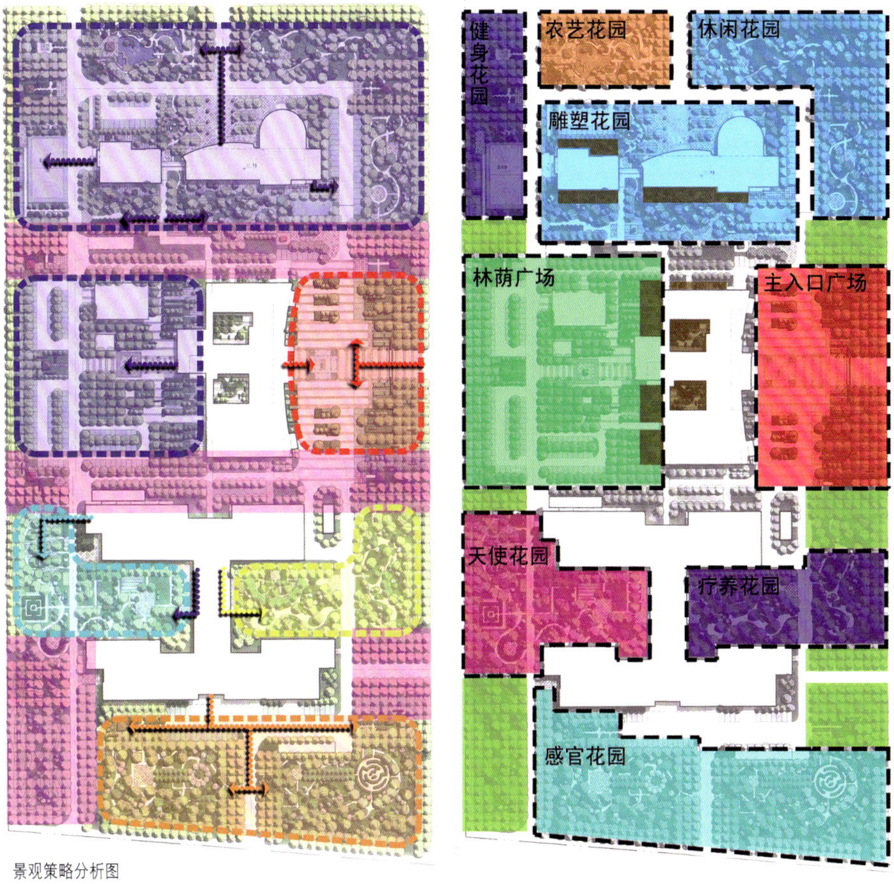

景观策略分析图

景观策略——相互隔离的“园中园”

现代医疗建筑趋向庭院化，这不但反映了人类向往自然的本性，而且具有一定的医疗价值。地坛医院的建筑布局就是典型的庭院方式。方案根据建筑的布局及出入口流线方向，提出了“园中园”的概念，即把整个医院作为一个大园林，根据建筑功能属性、出入口流线方向、建筑平面布局等要求，划分出若干个相互独立的主题花园。医务人员和不同类型的传染病人拥有自己独立的休闲花园。休闲花园之间利用植物、地形、道路、自动门进行隔离。

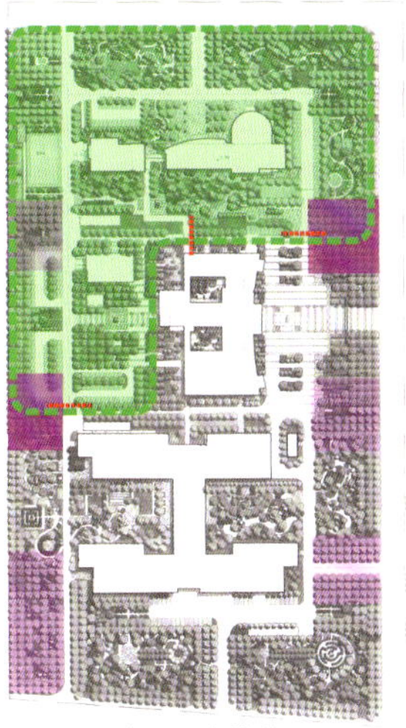

A级（绿色警戒）

通过关闭连廊及控制点内的伸缩门，保护办公区域不受污染

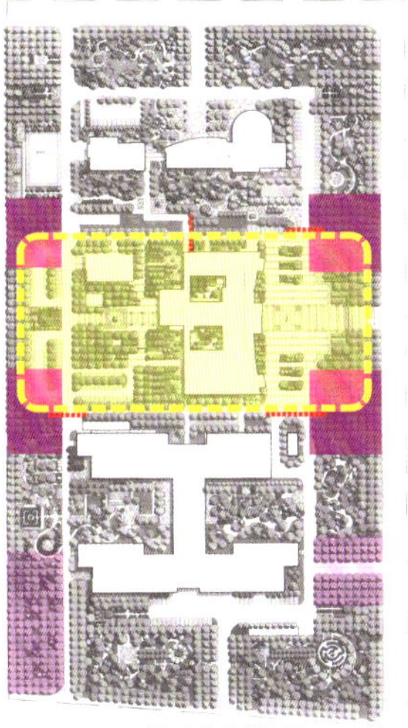

B级（黄色警戒）

将门诊部作为控制重点，将办公区、门诊部及住院部分隔开，减少交叉感染

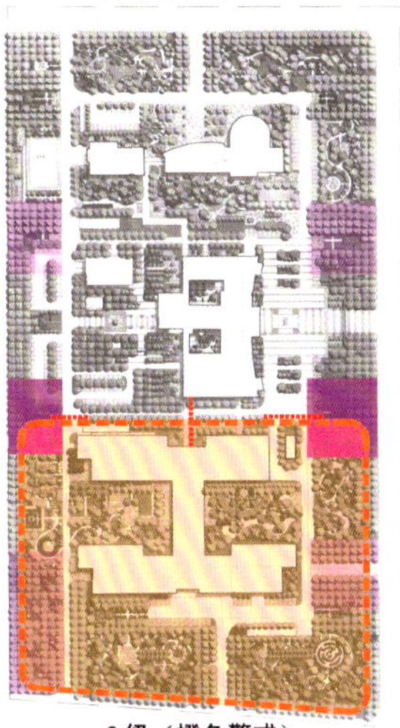

C级（橙色警戒）

严格控制住院部，保护其他区域不受污染

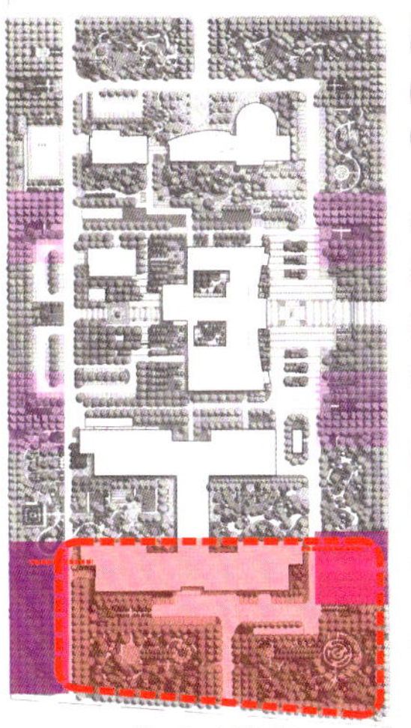

D级（红色警戒）

通过伸缩门等强制措施，严格控制呼吸科住院部，并将呼吸门诊临时迁至住院部，控制疫情恶化

管理策略分析图

管理策略——“平”“战”结合，四级警界防护隔离体系

根据传染病医院的特殊性，从总体到建筑再到环境贯彻切断传染链，控制传染源的核心原则。方案将“平”“战”结合的管理意念渗透到环境的设计理念中。利用四级警界防护隔离体系，使医院在日常的运营中可以发挥最大的效用，实现最低限度的交叉感染。当发生紧急情况时，四级警界防护隔离体系可以视情况开启，即可完全分开洁净区和污染区，切断传染链。

下沉广场效果图

Landscape plan and design of Beijing-Hangzhou Grand Canal-Tongzhou urban area section (first phase), Beijing

京杭大运河北京通州城市段（一期）景观设计

1

3

4

5

6

Styling and choice of the scenery ornaments. There are 8 light columns alongside both sides of the central square. The styling is derived from the sail boat ran in the canal and size of the columns is big enough to match the large square. During daytime, the 8 columns appear like sails floating on the canal, and the light in night seen in the water reflected the busy transportation in the past time. The ship anchor styling on the rails of canal side is derived from a dozen of anchors found in the canal side and now displayed in the Tongzhou Canal Museum.

FUNCTION ELEMENT

The key of the success of the project lies in how to create innovative environment. In today's Beijing, almost all riverside places have access for motor vehicles and the handling of the riverside space and greening is monotonous. The new project makes traffic paths away further from the bank and the greening space and pedestrian paths closer. And some areas adopt grassland riverside tactics, making the space full of changes.

The banks are the best places for the folk-custom events as well and people can further understand what have happened in the past time.

ECOLOGY ELEMENT

The project has taken full considerations of city ecology and protection of water resources. Some dry wells were designed to collect rain water. Many places have their best to adopt fewer pavement areas to maintain the existing trees and a greater amount of local plants were chosen too. The water sceneries in the square were designed as simple pools and the pavement riverbed helps make the water flat in summer times and they become pavement squares in winter times.

1. 水槽小品灯光设计
2. 广场帆形灯柱
3. 运河文化广场改造后
4. 以水槽为设计元素
5. 广场中央的千年运河文化步道
6. 生态滨水步道

1

工程档案

工程名称
京杭大运河北京通州城市段（一期）景观设计

工程地点
北京市通州区

设计单位
易兰（亚洲）规划设计事务所

设计师
陈跃中

项目景观总占地
170公顷

设计完成时间
2007年2月

竣工时间
2007年9月

2

3

4

1. 昔日帆影再现于现代城市景观
2.3. 栏杆小品的设计提炼于运河文化中的船锚
4. 帆型艺术灯柱夜景

Plan and design of Feicui Shancheng, Huizhou

惠州翡翠山城规划设计

N
01 奥斯克前广场
02 梦之岛喷泉广场
03 树阵广场
04 托罗姆瀑布湖
05 拉莫尔广场
06 后院阳光
07 迷你高尔夫场
08 眺湖平台
09 美斯德湖
10 优胜美地国家公园
11 内达华山脉休憩处二
12 半圆广场
13 入口广场
14 半山喷泉
15 西雅图广场
16 太阳广场
17 波特兰玫瑰湖
18 巴尔波阿湖
19 肯特公园
20 海洋之星眺台
21 团圆广场
22 美丽花园
23 达拉斯之冠
24 岩湖之光
25 尼里广场
26 沃斯堡花园
27 比尔街
28 奥斯卡后广场
29 加州花园
30 蓝色港湾

项目基地位于广东大亚湾经济技术开发区西区，为西部综合产业区中的服务区，是集购物、旅游、休闲、商住等于一体的高档社区。

设计构思

项目设计提出景观依山而造，凭借基地独特的地形优势，利用丰富的山、水资源，借鉴美国加州的造景经验，融合当地特色，以运动、休闲、养生为主，形成大的生命景观链。从山顶到入口的商业广场，融合建筑特点，成为翡翠山城的主景链；小区入口广场以水面为背景、商务会所为对景，使自然景观和人文景观有机融合；沿着景观链在不同的区段设不同的景观主题，并以雕塑、小品等构筑物形成对景；生态上则以植被为主，形成大的绿肺。

设计节点

商业设计

虽然商业不是社区的主要部分，但它是社区活力和健康发展的最直观表现。在商业街的设计中，我们充分利用建筑元素，通过材料、色彩、屋顶、流线、空间等表达设计概念，以此来创造本案的可识别性和别具特色的视觉及空间感受。

环境设计

环境设计引入阿尔卑斯山、北美加州或东南亚风情，不同的组团拥有不同的主题。建筑设计紧密结合环境，采用吊脚、错层、架空等形式来顺应山坡地形。环境规划充分利用山体及涌沟，有效组织利用地表雨水，结合保留树林，广植草皮、灌木、乔木等植被，与大绿洲的涓涓溪流及西边的河水融为一体。

水体设计

在区域内现有低洼处的湖面景观成为居住区可借用的水景，充分利用原有的大冲沟及小冲沟，将冲沟处适当改造成长年流动的潺潺小溪。以生态、趣味、科学为主题，根据不同地势上的水域，分别形成谷水、溪水和湖水三种类型。谷水以观赏性景观为主，溪水以趣味性景观为主，湖水则是亲水性景观为主。利用地形地势高差变化，穿插设计跌落有致的瀑布景观，汇聚到湖中。水体设计大小变化、线面结合、动静有别。湖面则结合公建布局改造成为社区园林露天泳池。

工程档案

工程名称
惠州翡翠山城规划设计

工程地点
广东省惠州市

设计单位
爱普斯顿国际

设计师
王艳华、Jaeger Johannes 、胡志高、卢东晴、濮新强、宋芸芸

项目景观总占地
71.5公顷

设计完成时间
2007年1月

竣工时间
在建

The project, designed as a service arm of the comprehensive western industrial zone, is located in the Dayawan economic and technology development zone of Guangdong province. It is designed as a high-end residential community for shopping, tourism, leisure activities and residence.

DESIGN CONCEPT

The sceneries are designed to be compatible with the hills, making use of the opulent mountain and water resources, adopting the concept in California State of the US and forming a large life scenery chain. The main scenery chain scatters from the hill top to the commercial square of the entrance. The business chamber and water background at the entrance square is a dynamic combination of the natural and human sceneries. Alongside the scenery chain are varied concepts, supported by reflected sceneries of sculptures and ornaments. A lot of plants help form a large green lung.

DESIGN SPOT

Commercial design

The shopping area is a direct reflection of an energetic and healthy community. The design makes full use of the factors of material, color, house roof, line and space, which are expected to create their identities and different visional and space feelings.

Environment design

The design introduces the elements from the Alps, California and southeast Asia for different complexes. Other mountain area elements include suspension, alternation and hollow. The design makes full use of the hills, waters, rain water, forests, grass, shrub and arbor.

Water design

The design alters the former lakes into water sceneries for the residents. There are three kinds of water sceneries: valley, stream and lake. Sightseeing, interesting and waterside sceneries are designed for these three categories. The lakes are redesigned as swimming pools for the residents.

Landscape planning design of Binhai international Plaza, Binzhou city

滨州渤海国际广场景观规划设计

现状分析

本地块位于滨州中心城区最繁华的商业街区，东临渤海七路，北至黄河四路，南至黄河三路，是滨州黄金地段。本次调整设计方案是在A区、D区已相继投入使用的情况下，根据市场需求做出的。将本地块作为整体来规划设计，通过对人流、车流、物流路线的合理规划和组织，对商业空间、商业形态及商业环境进行创造性设计，将本地块建成为集购物、文化、休闲、娱乐为一体的综合商业城。

设计构思

根据对滨州目前的商业调查和对将来的发展预测，本调整方案亦将遵循分期开发的原则。为了带动地块南面的商业人气、完善商业营运模式及商业形态，在地块的南面对原方案沿黄河三路及地块西面建筑作适当调整，沿黄河三路布置了21层星级酒店、大型高档百货商场；地块西侧布置八幢15～20层的高层住宅及在其裙房内设置规模相对较小的配套商业，以提升道路两面的商业人气。这是本次规划设计方案调整的重点。由原方案以中心广场区域为视觉中心，调整为以中心广场和黄河三路为视觉中心，提升了整个地块的城市景观形象。

整个建筑群体设计以现代的手法来处理。商业部分通过凹、凸、进、退、挑、透空、错位、退台多种手法创造出活泼多样的空间形态，立面有节奏地进行设计，通过体块的不同组合，材质肌理的变化，形成整体上的变化与统一。

技术说明

本次方案调整是在原有方案的基础上

THE CLEARANCE
FENDI

CUBA

Dior

TAITTINGER

IECOL'S

TAPISSIER
TAITTINGER

Panasonic

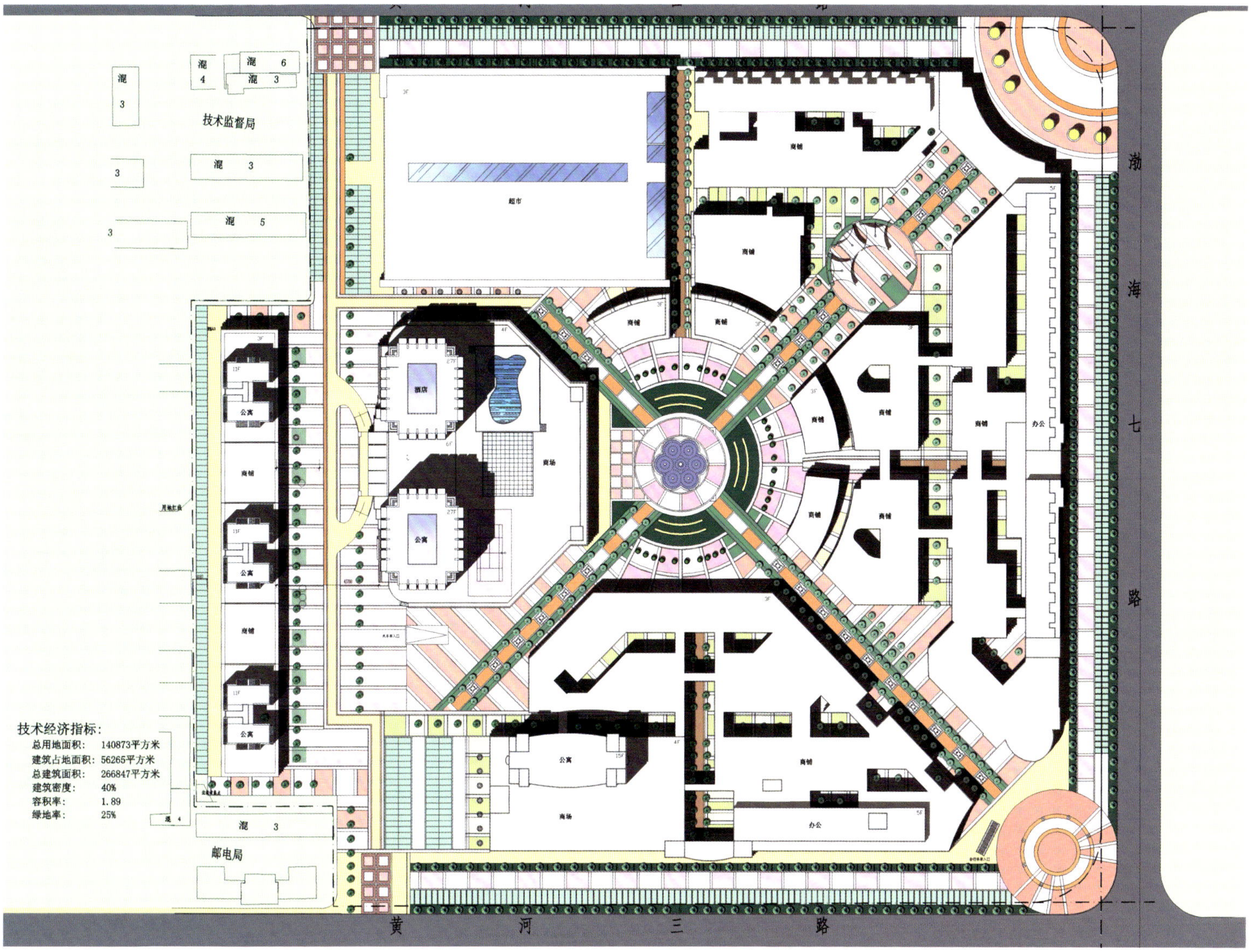

对空间形态进行精心设计，总体空间规划讲求大气势、大手笔，在小尺度的步行商业空间设计中又注意宜人的尺度和趣味性。广场是现代城市空间环境中最具公共性、最富艺术魅力、也最能反映现代都市文明和气氛的开放空间。通过对水景、灯光、雕塑、城市小品、绿化等的设计，使直径约110米的中心广场更加合理、完美，成为一个舒适宜人的购物休闲场所。

地块内绿化系统主要是静态的集中绿化、屋顶平台绿化与动态的街道系统绿化。强调绿化系统的整体性和生态性，开发绿化系统的空间性和有机性，将每一寸土地都利用起来。绿化设计主要集中在中央广场，由于广场尺度较大，缺乏庇荫空间会缺乏人气，因此广场四周采用乔木类植物为主，适当间隔，形成尺度宜人的休憩空间，并与面向广场的高层形成良好的景观氛围。其次，通过两条连接入口广场与中央广场的景观轴绿化，营造舒适宜人的休闲空间。另外，通过建筑围合的内街庭院及建筑退界产生绿化空间，形成绿化节点，其间设置草坪、花坛、雕塑小品、休闲座椅，创造立足自然的舒适空间。空中绿化主要在空中连廊和屋顶平台上展开，通过空中绿化削弱楼层感。

Landscape design of Liangshan Nationality Culture Centre and Torch Plaza, Xichang city

西昌凉山民族文化中心及火把广场景观设计

西昌市是凉山彝族自治州首府，四川省有名的历史文化名城，享有“月城”的美誉。旅游资源丰富，更有独具特色的彝族文化和每年一度的国际火把节盛会。火把广场的设计将满足城市大型活动及市民日常休闲活动的要求。

火把广场位于西昌市西南面，紧临邛海。用地为扇形平面，三边紧临城市道路。扇形中心线将用地分为东西两个部分，沿中心线上布置了一组三个广场。东侧布置了民族艺术中心，是一个斜插入山体的建筑；西侧布置文化休闲公园，为自然隆起的山体。位于中轴的三个次一级广场从北往南是柱阵广场、中心广场和树阵广场；在东区景观点有杆影广场，西区景观点有星坪、花轮和磨秋坪。

中心广场置于用地中心，圆形，广场半径60m，边缘距城市

1. 建筑与环境和谐统一
2. "永恒之火"与文化中心
3. 柱阵广场日景
4. 民族纹样成为独特的景观元素

1

路90m到120m，围合在高起的建筑和山体之间。广场中央为“永恒之火”雕塑，放于圆形静水池中，雕塑由国家美术大师韩美林先生创作。广场铺地选用当地出产的红色与灰色的花岗石，作不规则交错铺装，表现火的涌动与天体的运转。地面结合灯光处理，表现星宿的关系。圆形广场用草白玉收边，其间放置高出地面的节日石以及与地面相平的月历石，表现彝族传统的十月历和汉族的农历，并表示彝族和汉族的传统节日。广场外缘沿山体有多重环状草白玉石阶，石阶随山体升高，成为观演看台。

中心广场北侧的柱阵广场为入口广场，长度150m，宽度从33m到100m。中央

2

3

1. 火把广场鸟瞰图
2. 火把广场及文化中心平面图
3. "永恒之火"雕塑夜景
4. 火把广场实景鸟瞰图
5. 民族纹样是建筑表皮的灵感之源

4

5

工程档案

工程名称
西昌凉山民族文化中心及火把广场景观设计

工程地点
四川省西昌市

设计单位
中国建筑设计研究院环艺院景观所

设计师
崔恺、李存东、李力、张男、巩磊、雷洪强、史莹芳、姜蕾、陆柳、ERIC

项目景观总占地
15公顷

设计完成时间
2006年8月

竣工时间
2006年12月（一期建成），二期待建中

为黑虎雕塑，由艺术家武海鹰创作。广场后部成弧形布置56根直径2m的民族之柱，代表56个民族团结在一起，柱子选用当地产的红砂岩砌筑。柱上嵌有彝族和藏族纹样浮雕，浮雕由艺术家张宝贵创作。

火把广场南侧的树阵广场，长度140m，宽度与柱阵广场相当。广场地面的图案来自彝族传统纹样，两侧种植竹子，南侧植有挺拔的水杉。广场南端的花桥平缓升起，跨过城市道路，通向南侧的民族文化生态园。凉山之鹰雕塑从桥中高高探出。

整个火把广场围绕永恒之火雕塑旋转和舞蹈。

1. “水”与“火”交相辉映
2. 广场全景
3. 广场入口的黑虎雕塑

Xichang is the capital city of Liangshan Yi nationality autonomous region, one of the famous historical and cultural cities of Sichuan Province. It is also known as Yuecheng. Its tourism sources include its peculiar Yi nationality culture and once a year International Torch Festival. The Torch Plaza needs to be designed for large-size mass activities and daily leisure holdings of the residents.

The Torch Plaza is located in the southwest end of Xichang city, very close to Qionghai. The land is flat, like a fan, open to city roads in three sides. The fan central line divides the land into two parts-east and west. There are 3 plazas in one group alongside the central line. The east side one is a national art centre, a building cutting in the hill. The west side part is a culture park, a naturally ridged part of a hill. Alongside the central line are three sub-plazas: column plaza, center plaza and tree plaza, from north to the south. Other sites include the ganying plaza in the east area and Xingping, flower wheel and Moqiuping in the west area.

The round centre plaza is located in the centre of the land, surrounded by rising buildings and hills. Its radius is 60 meters and its edge is 90-120 meters away from the city road. There is a sculpture named Pertinent Fire in the plaza centre, put in the round pond. The sculpture was designed by Mr. Han Meilin, a master in the field. The floor is made by local-supplied red and grey granite bricks, which are arranged in an irregular manner to show the movement of fire and operation of the astronomy. The floor lights are designed to illustrate the constellation relationships. The round plaza is enclosed by white jade, with higher-than-the-level festival stones and month-calendar stones which are of

3

2

the same level with the floor being placed upon it. The stones are designed to represent the 10-month Yi lunar and the Han lunar calendars as well as their traditional festivals. Alongside the outer edge of the plaza, there are multi-end white jade steps, which rise up to the hill as spectator terrace.

The north end column plaza is one for entrance. The plaza is 150 meters long and 33-100 meters wide. There is black tiger sculpture in the centre, designed by Mr. Wuhaiying. In the backward end of the plaza, there are 56 columns with a diameter of 2 meters. The arc arranged columns represent China's 56 united nationalities. The columns are made from local-supplied red sand. Embossment of the Yi and Tibet characters was made on the column. The art was designed by Mr. Zhangbaogui.

The south end tree plaza is 140 meters long and its width is similar to the column plaza. The floor pattern was from the Yi tradition. Bamboo was planted alongside both sides of the plaza and there are metasequoia at the south end of the plaza. A flower bridge rises steadily in the south end of the plaza, crossing the city road, and leading to the south end traditional culture ecology park. A sculpture named Liangshan Eagle rises highly from the bridge.

1. 光与影的魅力
2. 广场夜景鸟瞰图
3. 乡土材料与乡土植物
4. 柱阵之中的震撼

Construction plan of the new campus of Liaoning Administrators College of Police and Justice

辽宁公安司法管理干部学院新校区建筑方案设计

景观规划

本案的绿地景观规划以点、线、面结合的方式组织绿地景观网络，以建筑界面依托山体围合场地，绿廊串接各个开放空间节点，各级空间分合有致。

依据场地的功能布局，将绿地景观分为四种类型：自然林地景观、林下广场景观、过渡带景观、滨水景观，形成满足不同功能需要的景观特色。

自然林地景观

包括校区主干道和水道外围的自然山地。景观设计以保护自然植被为基础，同时增加植被的多样性和层次性。林间以小路相通，采用生态化的道路护坡，间隔设置休息平台、座椅、景观小品与标识，营造自由、浪漫的林间氛围。

林下广场景观

包括入口广场、校史广场、中央广场和山顶广场，沿校区主轴线成序列排布。广场景观以层叠的山地地形为基本构成元素，铺地和树木沿不同的地形线展开。铺地选择混凝土与砖材相交替，树木选择高杆树木，留足林下活动空间，其间穿插矮墙、座椅，形成严谨而富有变化、可供休憩的校园礼仪性空间。

过渡带景观

在广场之间、地形变化较大的区域形成过渡带景观，其间间隔种植苇草与阔叶树，力图营造生态化的意境。

滨水景观

在水道两侧形成滨水景观，水岸种植水藻、芦苇等水生植物，在不破坏生态的基础上，间隔设置休息平台和栈桥，增强亲水性。同时强化横跨水道的公共走廊的景观效果，增加滨水空间的景观层次。

出入口景观设计

出入口设计应当结合景区风貌，不宜过于夸张。北侧出入口紧邻沈棋路，人流量大，应当布置展示学院形象的校门和入口广场，规模较大。南

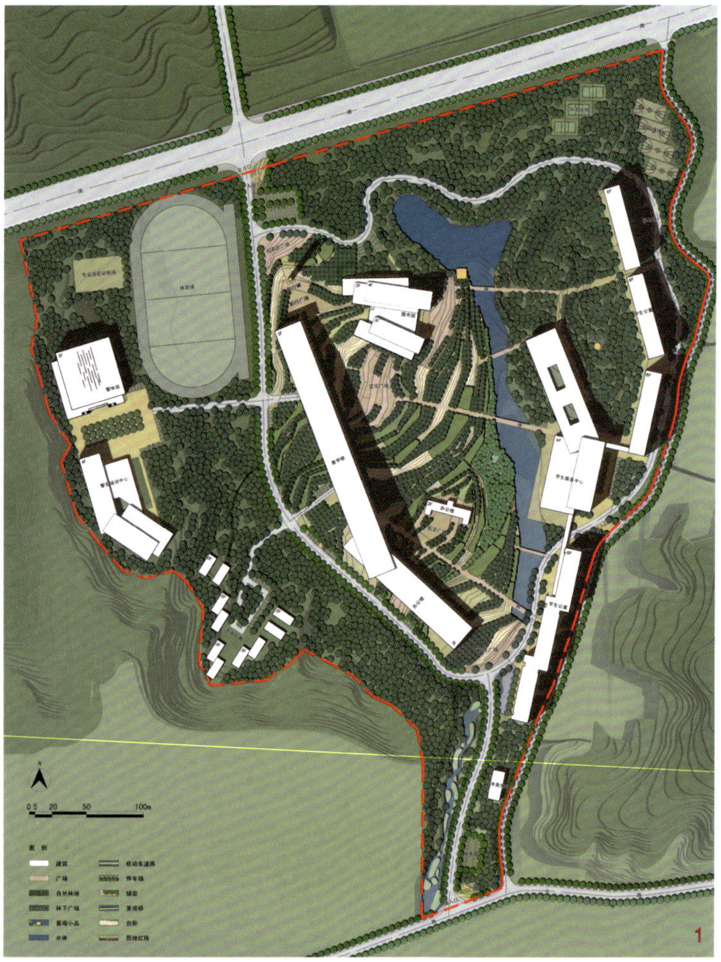

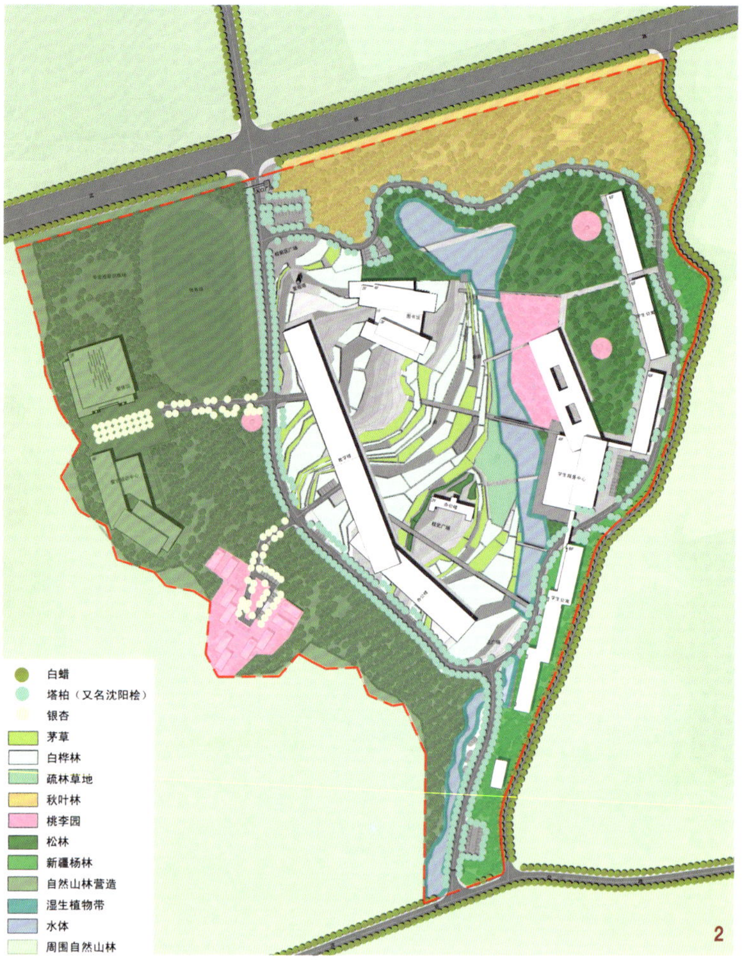

1. 总平面
2. 植物配置图
3. 景观结构分析图
4. 功能结构分析图
5. 绿地系统分析图
6. 道路交通分析图

侧出入口接入沈扶公路，现状场地绿化好，宜结合环境，设计园林式入口。

绿化类型及植物配置

道路绿化

外围行道树建议用白蜡，秋景观叶为主；园区行道树采用塔柏，以高大挺拔的姿态树立高昂正直的形象，与司法学院的整体形象相互呼应。

教学区植物景观营造

依地形和设计场地走势种植茅草和白桦林，并以其为主景，营造一种原生态、简洁、大方的植物景观，呼应整体建筑景观；水边开拓一块空旷的疏林草地，为课余活动、观景等提供空间。

生活区植物景观营造

桃李园：以山桃、京桃、碧桃等桃树为主景树，辅助稠李、山杏、油松、塔柏、云杉等树成片成林种植，寓意“桃李满天下”。

银杏林：秋景观叶、观果，做支路、广场、树阵等的景观节点，列植、丛植皆可。

松林：延续山脉种植，以渗透的方式进入园区，形成学习的一种背景——“松涛阵阵”，主要采用油松、云杉、冷杉等物种，间隙可种植一些秋叶树。

自然植物景观营造

片植新疆杨，与塔柏相辅相成，呈现昂扬向上的气势，营造具有震撼力的入口景观；以栎属（辽东栎、蒙古栎等 ）为主，成片成林种植秋叶林，可辅助种植五角枫、银杏、山桃、稠李、卫矛、红瑞木、火炬树、槭树科等观赏树木；延续山脉种植环境，以松林和刺槐、杨树种植为主，采用粗犷的种植形式营造出野生山林的生态环境。

湿生植物带景观营造

以芦苇、菖蒲种植为主，辅助种植垂柳、香蒲、千曲菜、荷花等，营造亲切、自然、生态的水岸环境。

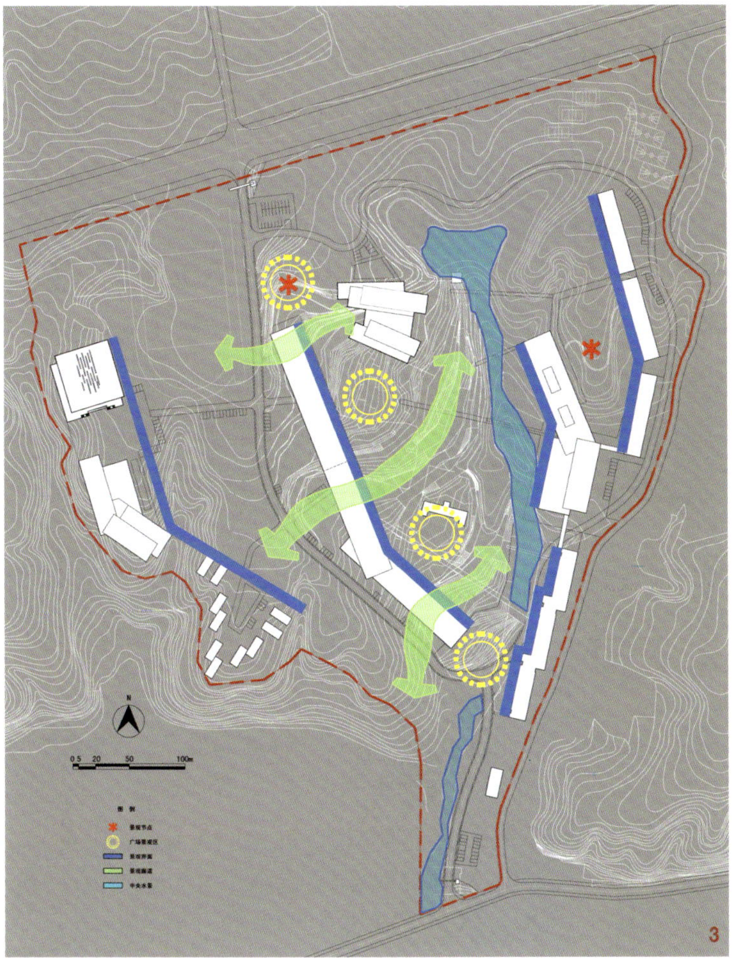
0 5 20 50 100m
图 例
景观节点
广场景观区
景观界面
景观廊道
中央水景
3

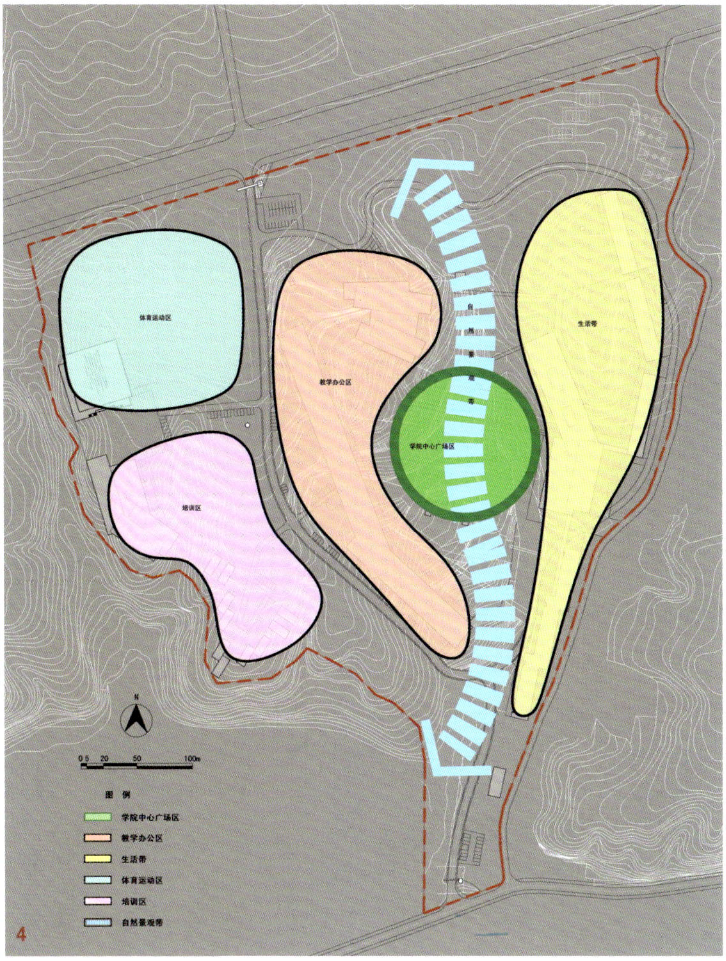
体育运动区
教学办公区
培训区
学院中心广场区
生活带
0 5 20 50 100m
图 例
学院中心广场区
教学办公区
生活带
体育运动区
培训区
自然景观带
4

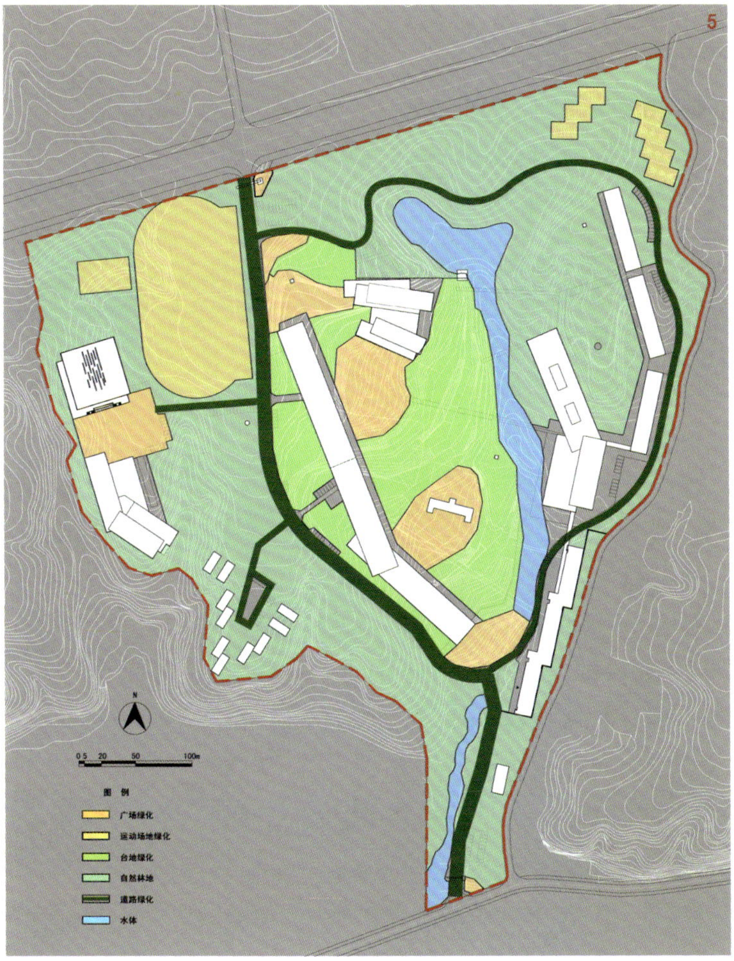
5
0 5 20 50 100m
图 例
广场绿化
运动场地绿化
台地绿化
自然林地
道路绿化
水体

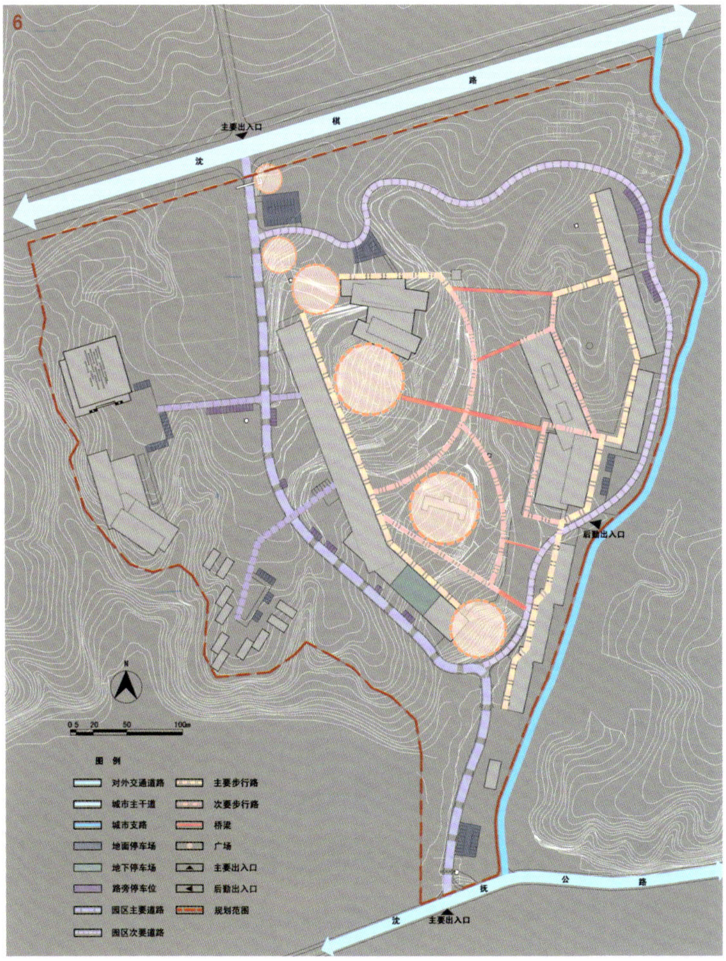
6
主要出入口
后勤出入口
主要出入口
0 5 20 50 100m
图 例
对外交通道路
城市主干道
城市支路
地面停车场
地下停车场
路旁停车位
园区主要道路
园区次要道路
主要步行路
次要步行路
桥梁
广场
主要出入口
后勤出入口
规划范围

1

2

SCENERY PLANNING

The design integrates scenery networks by means of point, line and facade.

The sceneries include 4 groups: forest, forest plaza, transitional belt and waterside sceneries.

Forest scenery

The area includes campus main paths and hills outside the water enclosing belt. The design is based on protecting the natural plants, making them multi ends and multi levels. The forest paths adopt eco road side slope protection works, supported by rest podiums, chairs, scenery ornaments and signs.

Forest plaza

The area includes plazas of entrance, college history, the central and hill top, being in order alongside the campus main axial line. The pavement is built of alternated concrete and bricks and there is enough shadow under high trees. And seat walls and chairs help make the place more charming.

Transitional area scenery

The transitional areas find use in between squares and areas that change greatly, supported by adaptable vegetation.

Riverside scenery

There are sceneries alongside riverside with bulrush and other kinds of hydrophytes being planted. The facilities include rest podiums and approach bridges.

Scenery design of the entrance area

The design needs to be modest. The entrance in the northern side should be larger as a displaying area and the entrance in the southern side should be a garden-like entrance.

3

1. 学生宿舍
2. 教学综合楼
3.6. 鸟瞰图
4. 警体馆
5. 生活服务中心

GREENING TYPES AND VEGETATION ARRANGEMENTS

Road side greening

The recommended spices are white wax and cypress which help improve the image of the college.

Vegetation scenery of the class area

The recommended spices are couch grass and birch which feature natural land, conciseness and decency. There should be a big area of grass land at the riverside for extra curriculum activities and other events.

Vegetation scenery of the dormitory area

Peach and plum garden, full of peach trees, supported by apricot, Chinese pine and spruce.

Apricot garden, serving as scenery points of sub-roads, squares and tree sections.

Pine tree garden, consisting of mainly Chinese pine, spruce and fir tree, as a background for the learning people.

Scenery planning of natural plants

The plants include Xinjiang poplar, apricot, David peach, plum and other spices to create a natural environment.

Scenery planning of wetland plants area

The recommended plants include bulrush, calamus, weeping willow, cattail and lotus.

1

工程档案

工程名称
辽宁公安司法管理干部学院
新校区建筑方案设计

工程地点
辽宁省沈阳市

设计单位
北京土人景观与建筑规划设计研究院

设计师
俞孔坚、文航舰、马特（德国）、林柏、
肖远、陈晨、杨虹桥、姜秀娟

项目景观总占地
24公顷

设计完成时间
2007年1月

竣工时间
在建

2

1. 专家公寓
2. 培训中心楼
3. 教学综合楼

商业街鸟瞰图

Environment design of the entrance area of Longqingxia landscape scenery, Beijing

北京龙庆峡景区入口区环境设计

1

龙庆峡风景区是京郊著名的游览胜地，景区以独特的自然山水地貌和植物景观为主，同时具有丰富和神秘的文化内涵；龙庆峡景区以保护森林生态系统和生物多样性、保护自然和文化遗产为主要功能，适当开展科学考察、风景游览、休闲健身、农业采摘等活动，是具有国家级风景区资源价值的省级风景名胜区。

本项目是整改项目，因此面临的问题相对具体。由于景区本身已经具有了相当的知名度，并且本身具有丰富的风景资源，因此本项目所面临的问题就集中在景区的进一步发展以及升级改造这两点上。

本方案将景区入口区分为三个部分：入口广场区、湿地公园区、商业服务区。各个区域都被赋予了不同的功能特色。

湿地公园区：该区以河道整治、岸坡防护与河道清障、保证行洪安全为基础，结合景观建设与生态修复，利用现有地势，改现有窄深直河道为宽浅曲河道，营造缓滩溪流，设置堆石跌坎，形成多级跌水，丰富水体景观，同时种植野生及观赏性湿生植物，恢复河道湿地生态系统，建设河滩地、自然林地、林地斑块、芦苇地、河坝和步栈道等，充分营造水环境。

入口广场区：在景区入口处设计建造滨水花园广场，同时对现有建筑进行升级改造，建筑立面进行统一整改，使整体建筑风格和谐统一。依照广场地形变化，采用形式感强烈的设计语言对广场的功能特性加以强化。在植物配置上注重高大乔木和地被植物的结合，营造丰富的林下空间，以打造绿色广场这一主题。

商业服务区：规划将景区大部分商业和餐饮设施移至该区域，集中管理。商业区建筑风格以中国北方传统建筑风格为主，以求更好的切合龙庆峡景区的整体形象。围绕商业点，在局部建设商业管理和服务设施，可以更好的为游客服务。

2

Longqingxia scenery area is famous in urban Beijing, with its peculiar hills, rivers and plants, in addition to the mysterious and opulent culture carriage. The Longqingxia area is designed to protect the diversification of the forestation and biology, as well as the protection of the nature and culture heritage. Other approved activities include scientific research, sightseeing, physical exercise, and agriculture.

The project needs to solve some detailed problems in improving the image. Due to its renowned status and opulent sources, problems it faces are upgrading and further development.

The project divides the entrance area into three parts: plaza, wetland park and commercial area with distinctive characters.

Wetland park. The area needs to improve the river bed, trim the hill and ensure flood river. The project makes use of the current deep and steep river bed and makes them wider, forging artificial beach and rivulet as well as stones and falls, in addition to multi-step water falls, while planting

3

1. 建筑立面
2. 建筑立面分析图
3. 总平面
4. 规划分区图

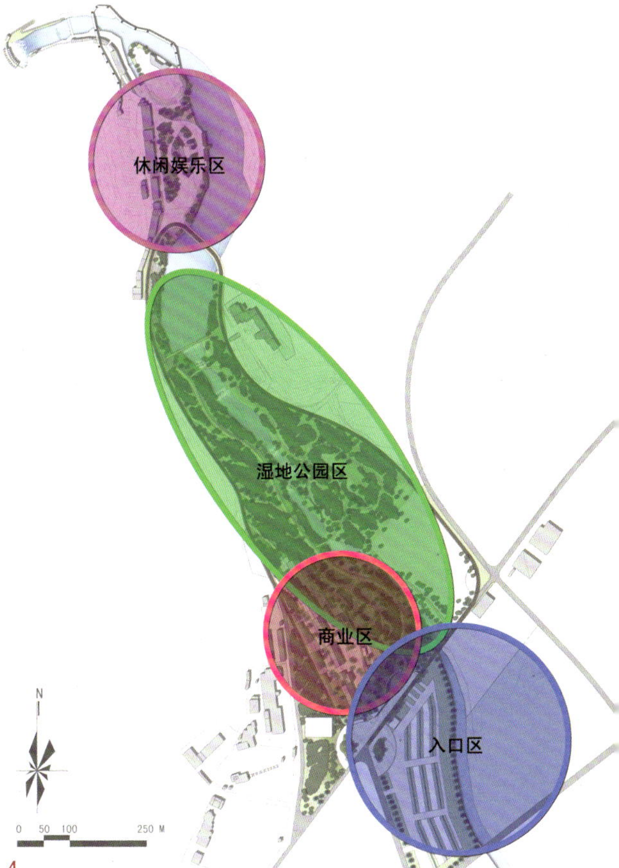

4

1

wildlife and scenic plants, to restore the wetland eco system. The designed tactics include riverside land, natural forest land, reed land, river bank and plank ways.

Entrance plaza. A waterside park was designed and current buildings need to be upgraded by trimming the facade. High arbor and grassland are combined to create a green plaza.

Commercial area. Most shopping and catering sites are expected to relocate here. The constructions will highlight the north style in China.

工程档案

工程名称
北京龙庆峡景区入口区环境设计

工程地点
北京市龙庆峡景区

设计单位
北京北林地景园林规划设计院

设计师
赵峰、周同、张大敏、赵铁楠、许建宇

项目景观总占地
36公顷

单位平米造价
270元

设计完成时间
2007年3月

竣工时间
在建

1.3.4.5 湿地公园
2. 商业街

2

3

4

5

Hi-space of Hot Spring & Leisure City, Beijing

北京“温都水城”之“水空间”景观设计

温都水城位于北京市昌平区北七家镇宏福生态社区内，地处平西王府旧址，正好在北京的中轴线上。其北面是唯一流经北京的天然河流——温榆河，西面是万亩生态园区和京昌高速路，东面是王府花园和汤立路，东南、西南两三公里处便是天通苑、回龙观两个全国最大的文化居住区，地理位置十分优越，交通便捷，环境优美，同时，丰富的温泉资源以及深厚的人文历史都是成功打造该项目的基础。

“水娱乐中心”是温都水城这一大型水文化娱乐综合体中的一颗璀璨明珠，其巨大的矩形体量、曲线形的屋顶和现代建筑风格在周围的中国古典建筑群体中十分醒目。“水娱乐中心”建筑面积约2万平方米，可以同时容纳近4000人，主要活动有戏水、游水、观水，是亚洲目前比较先进的以水为主题的水上娱乐场馆。

“水娱乐中心”平面为规整的矩形，由于水娱乐设备的要求，主体空间高大而单纯，北面局部为三层或四层。其建筑形式采用现代的手法，规整通透的玻璃幕墙加上波浪型屋顶，同时穿插一些色彩

1. 整体鸟瞰图
2. 水纹地面
3. 室外温泉庭院
4. 四合院中间行道夜景
5. 王府院入口夜景
6. 宾馆北入口前湖面景观
7. 两大区域衔接处

1

2

3

4

5

6

7

明快的体块。建筑中的水空间包含了室内和南向的室外庭院部分，立面采用点式玻璃幕墙使建筑更加通透，将室内和室外的景观融为一体。波浪型屋顶北高南低，为的是和南面较低的酒店、养生会馆和四合院等古建筑群在建筑尺度上协调呼应，在北面向上隆起，对北面的大型停车场形成控制性高度，远远看去仿佛手搭凉棚远望的主人在期待客人们的到来。

水空间的景观设计包括室内和室外两部分。

室内景观分有三个区：东区、西区和中央区。东区以造浪池和人造沙滩为主，主要营造出开阔通畅的感觉。浅色人造沙滩、深色木地板休息区等硬质铺地配以高大的棕榈树以及低矮的蒲葵、八角金盘等绿色植物，使主要空间在视线上基本没有遮拦。人们的视线越过低矮的植物可看到室外泳池和景观绿化，增加了空间的延伸感和连续性。造浪池蓝色“海”面的东西两侧是白色曲线防浪墙，既增加了立面的趣味性，又契合了造浪功能的需要，令人遐想。西区是以漂流河环绕的岛型和半岛型区域，包括滑板冲浪、儿童戏水池、互动水屋、天网、鳄鱼

1. 王府院主入口全夜景
2. 四合院内庭院景观一
3. 四合院平面图
4. 四合院俯瞰
5. 养生会馆功能池组区
6. 四合院内庭院景观二
7. 四合院内庭院景观三
8. 养生会馆泳池、功能池

1

2

3

4

池等主要区域，其间通过6座造型各异的景观桥相联或与中央区联系。东区由于功能较多，占地面积较小，在景观设计中力求营造出步移景移，别有洞天的氛围。漂流河采用自然石与木桩为河岸，一些钢柱采用缠麻处理，这些自然材料与西区整体的氛围相协调。中央区是更衣后进入室内时正对着的一小片区域，是东区西区和入口之间的缓冲区域，其间布置了9棵高大的霸王棕，树池中设置各种绿植花卉，树池外围安排了休息座椅。沿西侧防浪墙设计了两个小型的儿童池，既利用了空间，又便于家长看护池中嬉戏的幼儿。

室外景观主要分南北两区。北区为大型停车场和主入口广场，南区为室外休闲戏水区。

水娱乐中心北广场以"水"主题为核心。整个广场可停车近500辆，设计除考虑停车外还考虑到将来举行大型演出、集会等多功能需要，所以没有种植高大的乔木，而以草地和彩色有肌理的铺地为主。地面采用Bomanite彩色地坪，水网纹单元肌理，蓝色飘带图案既呼应了建筑屋顶形式，又如同在地面上翻卷起一波波海浪。入口广场沿建筑长向设置了140米长的水舞音乐喷泉，随着音乐起伏，水柱时而喷射高达12米，时而在人们脚边欢快涌动，加上空中的水雾造成的彩虹以及波光粼粼的地面，使人们在进入水娱乐中心之前就置身于"水"的世界。而入口广场的跳泉拱廊、喷水巨鲸、跌水水池均强化了这一效果，加上夜晚立面上以"跳跃的雨滴"为主题的霓虹灯、螺旋水纹般一波又一波流动的入口照明以及入口两侧的喷火柱，使广场每时每刻都能散发着迷人的魅力。

水娱乐中心南面的室外休闲戏水区，是水娱乐中心戏水活动的一部分，是室内水娱乐空间的延续，它包括室外泳池、儿童池、温泉泡池等，在景观设计中将其与室内泳池、漂流河一并设计，相互融通，浑然一体。其室外泳池采用自然曲线形造型，水面围绕的阳光岛上有

6

5

张拉膜伞盖和休闲躺椅，是日光浴的理想场所，西侧设置水中汀步，供儿童玩耍。泳池南岸有林中温泉泡池，可边泡温泉边欣赏周围水景以及绿树蓝天。西侧漂流河室外区计划放置6头大小不一的喷水象，向漂流河和儿童池中喷水，带给人们一份惊喜。东侧是半圆阳光浴区与室外草裙舞演出舞台，以东面连接水娱乐中心和理疗中心的连廊为背景，加上各色植物配置，使人仿佛置身南太平洋的滨海度假胜地。

1. 岸边绿化
2. 彩色沥青路
3. 广场海洋主题
4. 广场波浪主题
5. 室外韵律围墙
6. 水中浮球

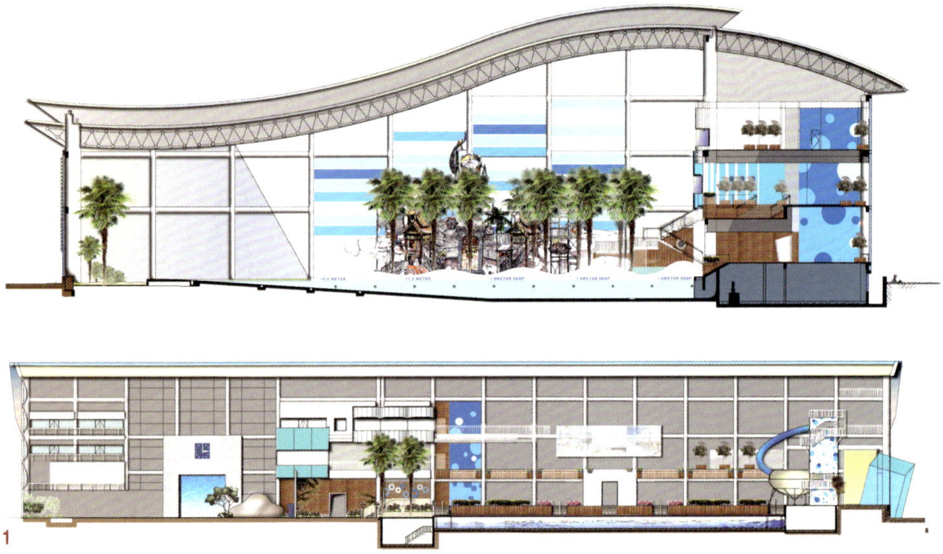

1

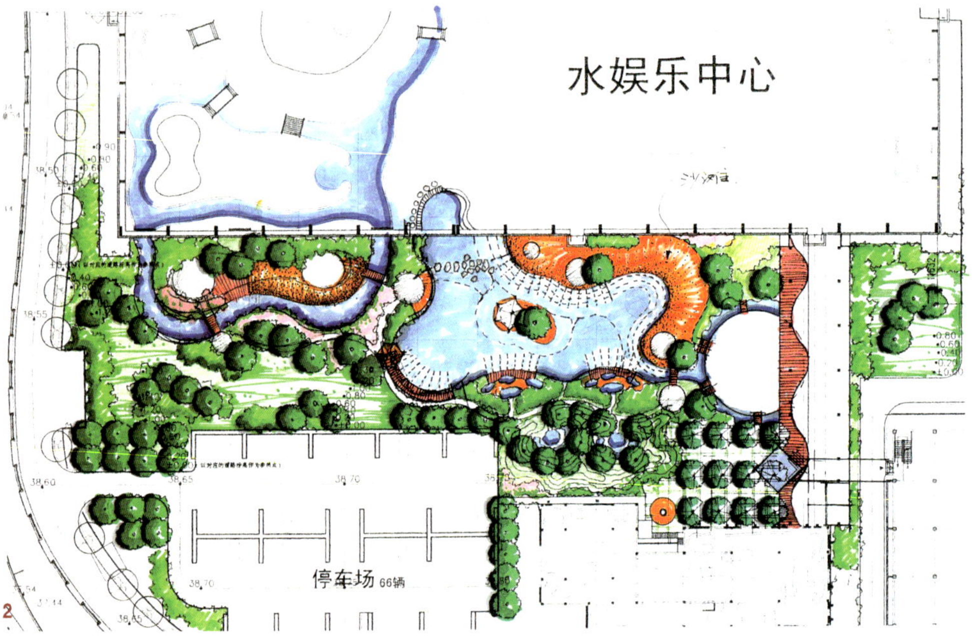

2

3

4

工程档案

工程名称
北京“温都水城”之“水空间”景观设计

工程地点
北京市昌平区

设计单位
澳大利亚SDG设计集团

设计师
聂建鑫、陈向清、叶伟聪、余向锋、翟莹、欧阳炜

项目景观总占地
3公顷

单位平米造价
500元

设计完成时间
2006年5月

竣工时间
2006年7月

Wendushuicheng is located in the Hongfu eco community of Beiqijia Town of Changping dist, Beijing. The place, used to be a prince home, is on the central axis of Beijing city. The land neighboring the Wenyu River, Beijing-Changping expressway, Tangli road and other residential communities in its 4 sides.

The Water Entertainment Centre features a huge rectangle mass, curved roof and modern building style. The 20,000 square meters building can house up to 4,000 people for entertainment, swimming and other events.

The water entertainment centre demands huge and simple space. Its northern part is 3 or 4 stories and built by modern approaches: glass curtain wall and wave-shaped roof, interlude by colorful masses. The water space consists of indoor and outdoor courtyard part in the south end. Its facade adopts glass curtain wall also, making the outdoor and inside sceneries integrated.

The indoor sceneries consist of 3 parts: east, west and middle areas. The east part mainly includes artificial wave pond and sand beach. All facilities make the area transparent in viewpoint. People inside can see the outdoor swimming pool and greening area, making the space sustainable and continuous. Alongside both sides of the wave pool are white color wave protection walls.

The west area are the island and peninsula areas, surrounded by drift current and 6 bridges are connecting them to the central area. The designers try to make the sceneries follow people's walk. The central area is a small area facing the inside door after changing clothes, also a transition of the east/west area and the entrance.

The outdoor sceneries are divided into south and north parts, for car parking and main entrance plaza and outdoor water entertainment respectively.

The parking lot can allow up to 500 cars and it is designed for the purposes of singing performance and collections as well.

Alongside the entrance plaza are 140 meters long water music fountain, with water column erecting up to 12 meters high.

The outdoor water entertainment area is part of the indoor activities. The swimming pool adopts the shape of natural curve. It is a perfect place for enjoying sunshine.

At the south bank of the swimming pool is hot spring pond and 6 water spray elephants will be launched there for children's entertainment.

1. 水娱中心西南剖面图
2. 水娱中心室外景观平面图
3. 王府院北立面夜景
4. 王府院内景
5. 入口
6. 互动水屋
7. 桥
8. 室外景区
9.12. 室外泳池
10. 开启屋顶
11. 室内漂流河
13. 室外泳池中小岛
14. 水娱乐中心

5

6

7

8

9

10

11

12

13

14

Landscape design of east bank of the canal, Hangzhou

杭州运河东岸景观设计

作为整个杭州运河旅游线的启动区块，运河东岸景观设计围绕运河文化旅游线展开，既挖掘历史文化底蕴，同时又满足新时代的需求。以已建成的运河文化广场为主体，强调景观"认知"特色，使游客能更全面深入地了解运河的历史。设计主要分为以下两个标段。

青莎古镇

以青莎古镇为主体，再现传统滨水古镇风貌。青莎古镇本是依托杭州运河北新关而发展起来的运河小镇，著名古建和园林学家陈从周先生就出生在这个小镇上。在其散文《故居》中，陈先生描述到："父亲自立后在杭州城北青莎镇散花滩建造了房子，我出生在这里。散花滩又名仓基上，可能南宋时为藏粮之处，四面环水，有三座桥通市上，三洞的华光桥，一洞的黑桥，还有一座叫宝庆桥。"整个景观设计构景依据陈从周先生的描述，进行了引用和提炼。引水入岸围绕成一个三面环水的亲水平台即"北新关广场"，广场上树一对象征北新关的标志性桅杆，平台通过两座桥梁和城市道路及南面沿河绿化带相连。作为主入口的桥梁一座取名"华光桥"，另一座取名"宝庆桥"。同时该平台也是整个青莎镇的主要入口广场，广场周围以樱花和桃花林结合卵石滩体现"散花滩"的意境。根据原有民居建筑的造型进行原地重建，整合成传统民居院落和街巷空间特色浓郁的步行街区，可引入"天禄堂国药""仁号南货""同福泰"等老字号特色店铺和"尚德堂""水竹居"等配套服务建筑，还可设置北新关历史展馆等文化娱乐设施。除此之外，还可利用古街和广场开展庙会、放河灯、文化节等活动。

1. 沿运河立面图
2. 沿运河透视效果图

1

2

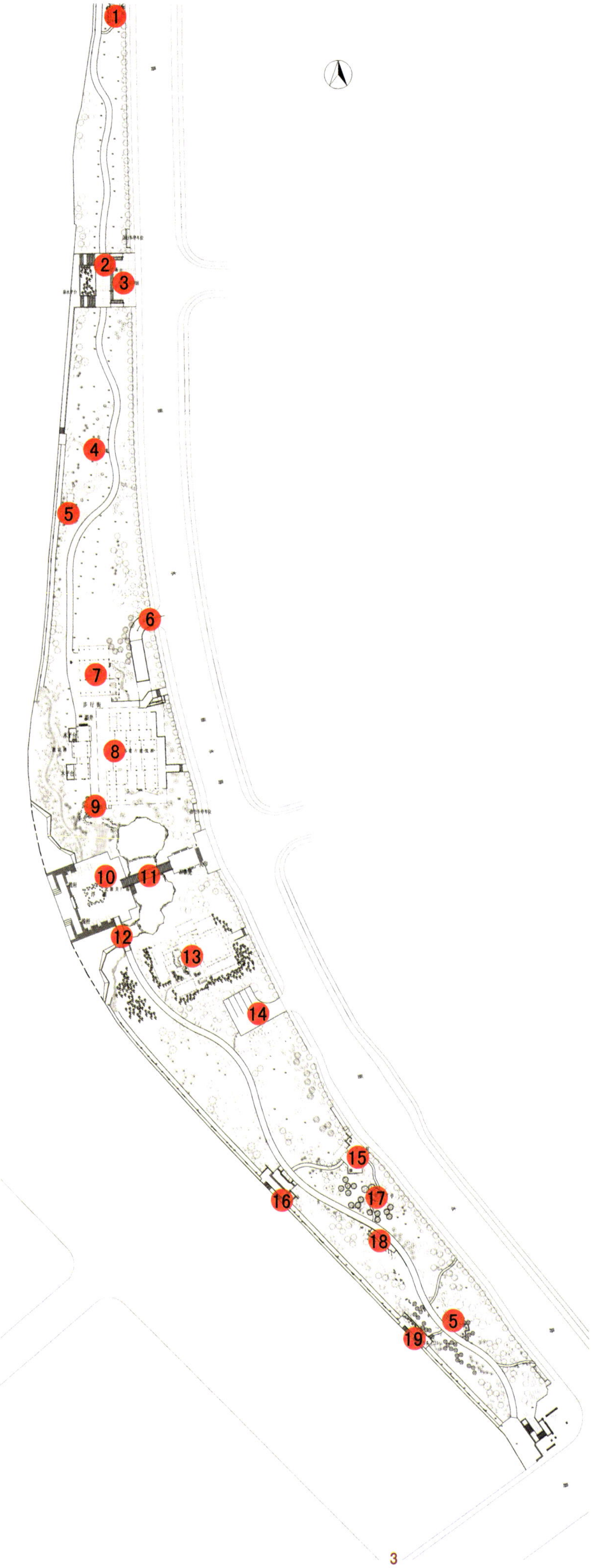

3

1.4. 复建古民居实景照片
2. 深巷酒香小品实景照片
3. 清莎镇平面图
5. 清可轩实景照片
6. 青莎公园沿丽水路侧效果台

1

As a key part of Hangzhou canal tourism project, the east bank scenery design focuses on exploring the historical details while satisfying modern life. The completed canal culture plaza is designed to make visitors acquainted with the history. The design has two phases:

1. 北新广场沿河实景照片
2. 北新广场沿道路入口实景照片
3. 宝庆桥实景照片
4. 船行平台实景照片
5. 拴船石及亲水平台
6. 北新关广场断面图
7. 桐心缘效果图

QINGSHA TOWN

The entire scenery design is based on the descriptions on the town by an expert. The Beixinguan plaza is in fact a platform with its 3 sides being surrounded by water. The platform connects urban roads and green belts in the south end through two bridges. The platform is meanwhile the main trance plaza of the town.

The previous residential houses were rebuilt. The pedestrian area plans to introduce famous traditional brands. Meanwhile, the streets and plaza can be used to hold mass activities.

2

3

4

5

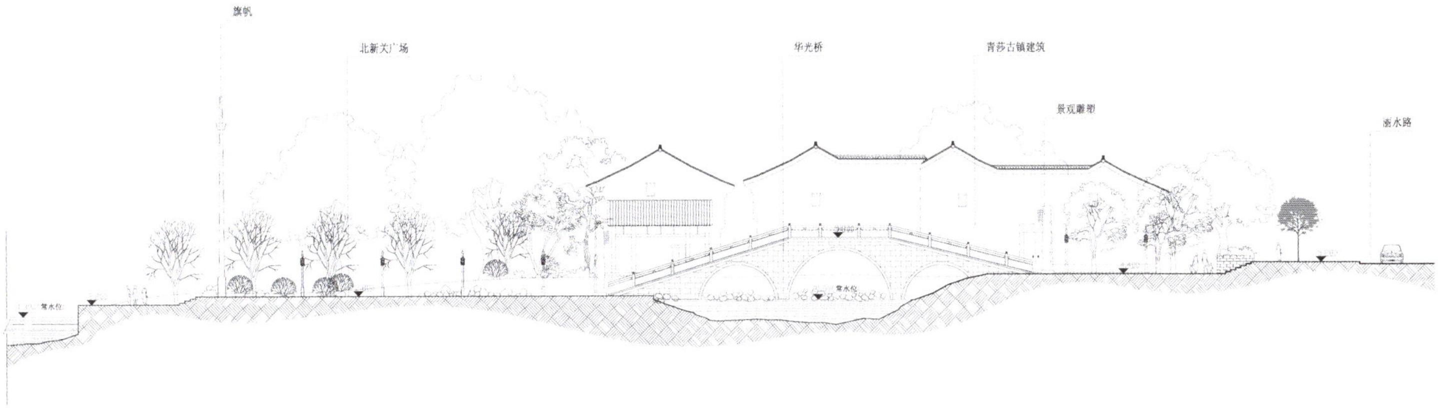

6

7

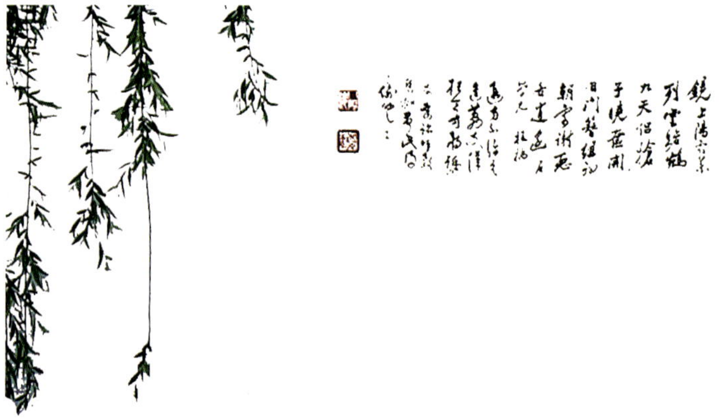

1

2

北星公园

北星公园是整个运河东岸北端的核心开放空间，公园以新建茶室、水上船坊等休闲设施为景观主体，结合运河滨水空间，行人可以漫步在河边别致的步行道上，也可以独坐岸边小憩，自然的岩石结合不经意的雕塑沿水岸随意放置，唤起人们对古运河历史变迁的回忆。

BEIXING PARK

The park is the key open space of the north end of the east bank, full of newly established tea houses and other entertainment facilities. The natural stones and sculptures remind people of the past times.

1. 北星公园沿河效果图
2. 北星公园鸟瞰效果图

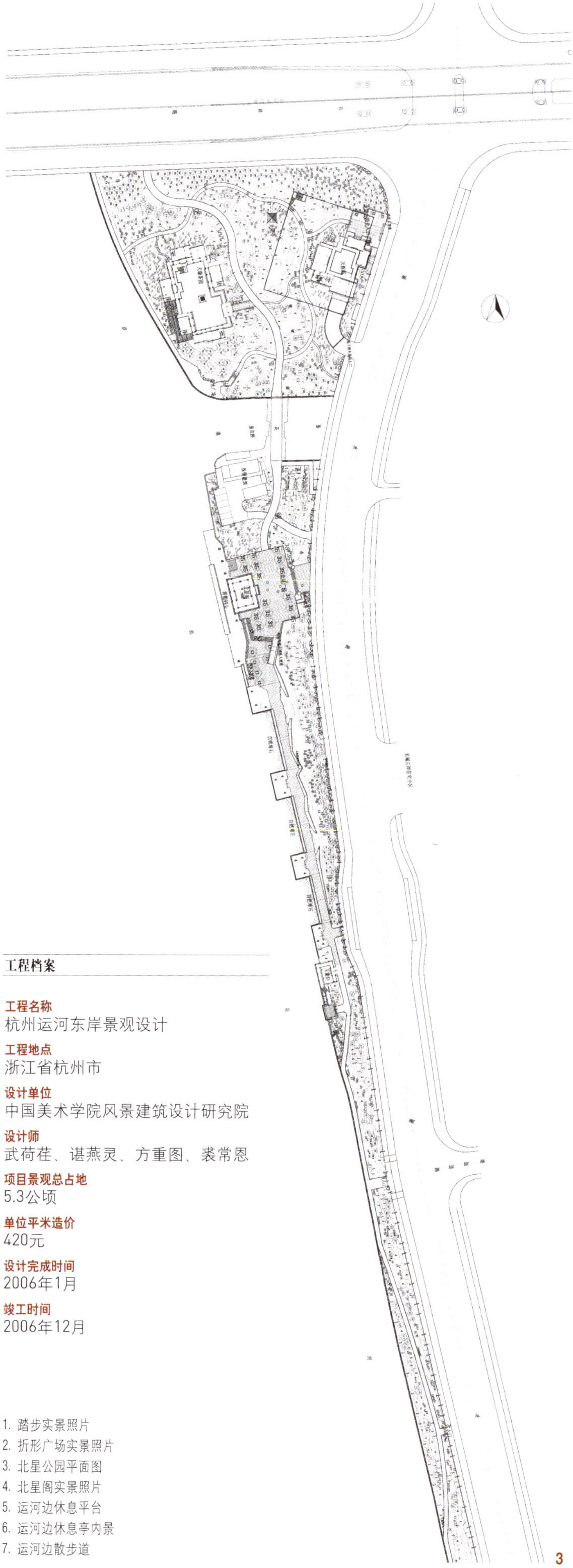

工程档案

工程名称
杭州运河东岸景观设计

工程地点
浙江省杭州市

设计单位
中国美术学院风景建筑设计研究院

设计师
武荷茬、谌燕灵、方重图、裘常恩

项目景观总占地
5.3公顷

单位平米造价
420元

设计完成时间
2006年1月

竣工时间
2006年12月

1. 踏步实景照片
2. 折形广场实景照片
3. 北星公园平面图
4. 北星阁实景照片
5. 运河边休息平台
6. 运河边休息亭内景
7. 运河边散步道

4

5

6

7

Landscape design of Qiandaohu hot spring village, Hangzhou

杭州千岛湖温泉度假村景观设计

项目概述

千岛湖松芦特色温泉度假村有着得天独厚的景观条件。地块南向面湖，东、西、北三面环山，北高南低呈正方形，从水面至峰顶高差约50m～80m，并与东、西、北三面分水线以内的集水面形成一完整的小流域，境内均被繁茂的天然林覆盖，附近已开发的特色温泉资源丰富，堪称一方风水宝地。

项目布局

入口接待区

主要由停车场、入口小广场和接待区三部分组成。入口区设停车场，可限制一些大型巴士进入度假村内。入口标识既有引导的功能，也是展示的窗口，设计用不加雕琢的块石挡墙组成，与自然环境浑然一体。接待区入口也设置简洁的引导标志，标志后种植青青翠竹，尽显雅致。

临水休闲会所

会所集亲水休闲、美食娱乐、室内SPA等多功能为一体，并可作为日后开展水上项目的服务性场所。内聚的建筑布局犹如“天井”，在“天井”的中心布置了“泉眼”——特色温泉。内庭空间可做一些露天吧台，满足人们休闲需求。会所临水的一面设计层叠的平台向水面拓展，并设游船码头，在日后的开发中可逐步开展水上游乐项目。

联体特色温泉会所区

360度的景观视角，高档的休闲设施和露天特色温泉，形成独特的康体特色温泉会所概念。会所努力营造家庭式的舒适内部环境，让人们尽

私人会所
(A型)
坡地别墅
(A型)
度假排屋
(C型)
联体别墅
(B型)
会所
经济技术指标：
规划用地面积：19992.00m²
建筑基底面积：4577m²
建筑总面积：10800.00m²
其中：1）A型别墅3栋建筑面积：2409m²
2）B型连体别墅建筑面积：2712m²
3）C型连体别墅建筑面积：2300m²
4）配套用房建筑面积：256m²
5）会所建筑面积：3123m²
建筑密度＝建筑基底面积：4007m² / 规划用地面积：19992.00m² X100% ＝20%
容积率＝建筑总面积：10800.00m² / 规划用地面积：19992.00m² ＝0.54

情享受在露天特色温泉泡池的高品质生活。

休闲假日特色温泉连排会所区

这里是与大自然零距离接触的休闲场所，可提供种类各异的特色温泉体验，如足浴、药浴、沙浴、泥疗、鱼疗特色温泉等。

山地特色温泉度假会所区

高档特色温泉概念的坡地会所，是高端消费人群的尊贵体验场所，是露天特色温泉SPA的超级享受的场所。豪华的私人会所，具有极佳的地理位置，远眺湖景背倚青山，超大的露台，放置豪华SPA温泉池，在大自然的怀抱中沐浴特色温泉。

生态体验区

保留区域内的原生态植被，以"少即是多"的开发理念为指导，强调"森林浴"的概念。野趣的木桩游步道依势而设，曲径通幽；爬山木栈道凌空铺设，直通生态观景台，凭栏远眺，美景尽收眼底。此区块作为一个中心的生态景观过渡区，具有自然野趣的特色，开发时应最大限度的保护原有资源。

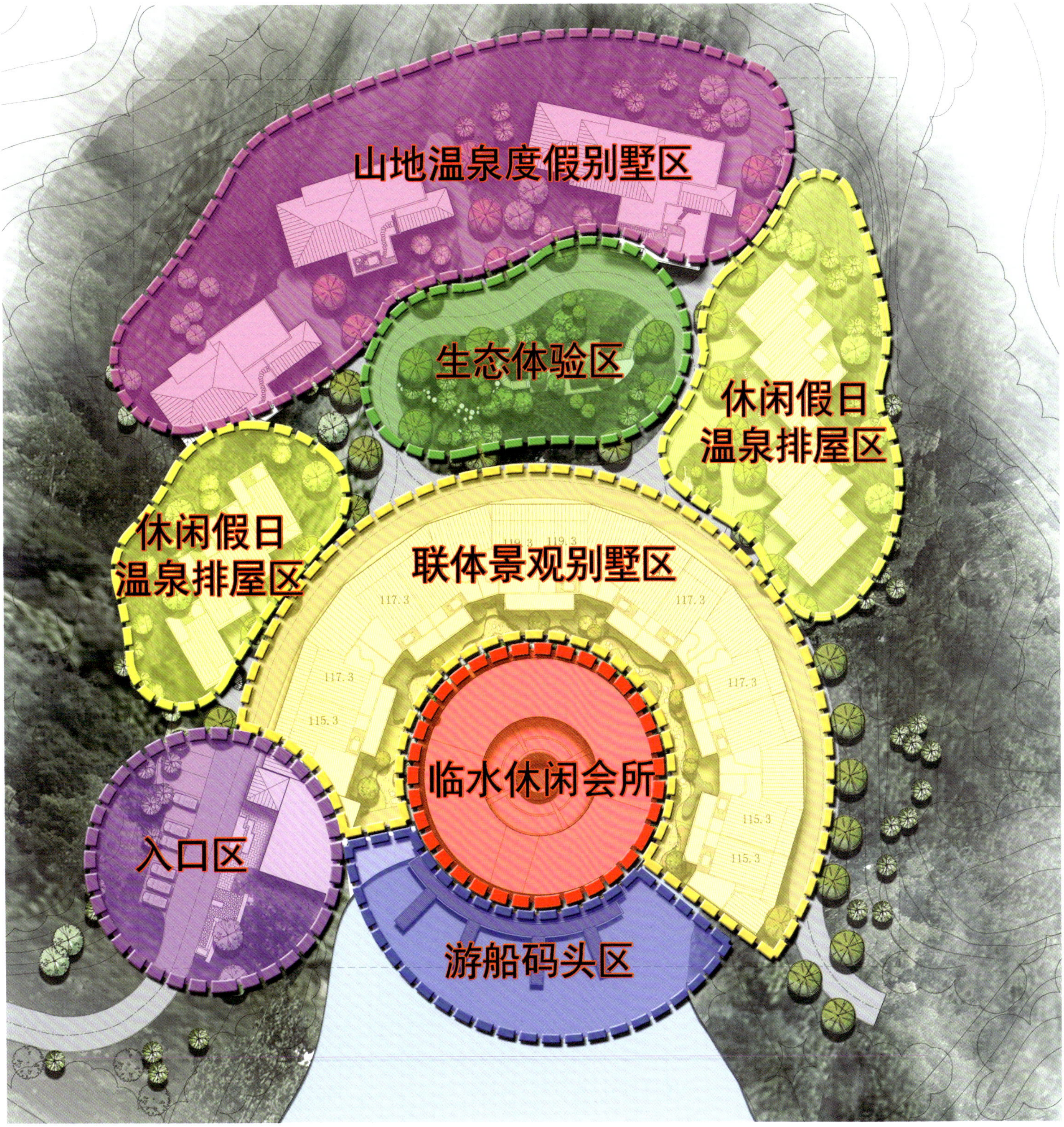

山地温泉度假别墅区
生态体验区
休闲假日
温泉排屋区
休闲假日
温泉排屋区
联体景观别墅区
119.3
119.3
117.3
117.3
117.3
117.3
115.3
115.3
115.3
临水休闲会所
入口区
游船码头区

工程档案

工程名称
杭州千岛湖温泉度假村景观设计

工程地点
浙江省淳安县

设计单位
杭州现代机构

设计师
郭宏强、傅晶、詹翰、方秋喜、赵凤花、杨春

项目景观总占地
8.29公顷

单位平米造价
300元

设计完成时间
2007年10月

竣工时间
2008年11月

PROJECT BRIEF

The hot spring village has peculiar conditions. The land faces hills in its east, west and north sides, and a lake in its south side. The square land is high in the north side and low in the south. The altitude from the water level to the hill peak is 50~80 meters. The land is covered by thick and natural forests.

PROJECT LAYOUT

Entrance reception area

The area includes a parking lot, a plaza and a reception area. Buses are not allowed inside the village. The entrance signs are made by naturally curved stones. Plants in the reception area makes it elegant.

Waterside chamber

The chamber integrates cuisine, entertainment and indoor SPA services. Buildings inside shape like a dooryard that houses some outdoor bars. The waterside of the chamber extends towards the water and a pier can serve water-based entertainment activities.

Hot spring chamber area

The tactics include 360 degree scenery view angle, high-end facilities and hot spring, in addition to family-oriented inner environment.

Hot spring bathing area

The section offers various kinds of hot spring baths.

The vocation and chamber area

The area offers services for those who want to spend a special period of time here.

Eco-experiencing area

The area offers services of forest bath and all tactics are built based on the original layout of the nature.

Landscape planning and design of LihuGuanshe area, Wuxi

无锡蠡湖管社山地区景观规划设计

基地解读

“蠡湖”因范蠡与西施的故事而得名。历史上，这里曾留下许多美丽的传说。整个区块内的每一处古迹都散发着浓郁的人文气息，也正如此，使得建筑有了生命力。因此，在保护这些历史建筑的同时，挖掘其人文内涵是我们规划这个地块的重点。

规划内容

本规划是从整体出发，关注整个蠡湖风景区的景观及功能，而不单纯着眼于规划区本身。综合分析周边及区块内的历史、人文和现代元素，创建一个以“太湖渔村”为主题、多功能复合的生态景观综合体，力求历史、人文与现代相融合。规划涉及的几个方面有：保留和完善万顷堂、驻美亭、镇湖庵、杨家祠堂、听涛园等具有一定人文历史的建筑；恢复和增加人文历史建筑，如翠声阁、观景长廊等；改造梅园水厂等部分建筑；开发建设原汁原味的太湖渔村聚落；开发建设游艇俱乐部等水上运动场所及美容、健身、饮食等休闲场所。

节点设计

历史文化区

整个地块保留建筑沿管社山东南麓散点布置，规划时将其合理组织、串联成为重点。沿湖规划一级规划道路，局部山腰地势平坦处，规划了一些休息场地。由于

万顷堂和驻美亭在沿湖立面上过于孤立，因此引入院落围墙，虚化建筑实体，刻划一种探访管社山庄故址的痕迹。

工业复兴区

将梅园水厂的部分建筑进行改造。如将四个净水池改造后埋于地下，直径都在20米左右，深度有3～4m，从地面上看，只有四池清水和一条工作通道。

蠡乡风情区

东管社村，以一个个院落为独立单元，依山面湖而建，整个村落呈自然环形放射状布局，生活在这里的人们，以打鱼为生。设计延续"渔村"的概念，对原来的东管社村的聚落肌理、巷道肌理、院落肌理通过批判的重构策略，创造性的将中国的传统建筑语言和现代手段和谐地融为一体，极力保持其原有的韵味以及其与山水之间的和谐关系,节约了开发成本。

水上运动区

本区是一个相对独立的区块，是一处高级会馆形式的综合娱乐区，主要开展以水上运动为主题的项目，包括水上运动培训、水上运动服务、美容健身、宾馆餐饮等。建筑多单层，局部二至三层，弱化建筑造型，强调建筑各单体之间、建筑群体与环境的和谐关系，整个建筑群掩映在绿树花丛之中。建筑材料以玻璃、石材、钢架为主，以灰白色为基调，以大玻璃落地窗、大面积石墙形成通透、开敞、室内与室外相融的建筑群。

锦
园
路
梁
溪
河
渤
公
岛
太
湖
N
0 20 40 80M

总平面图

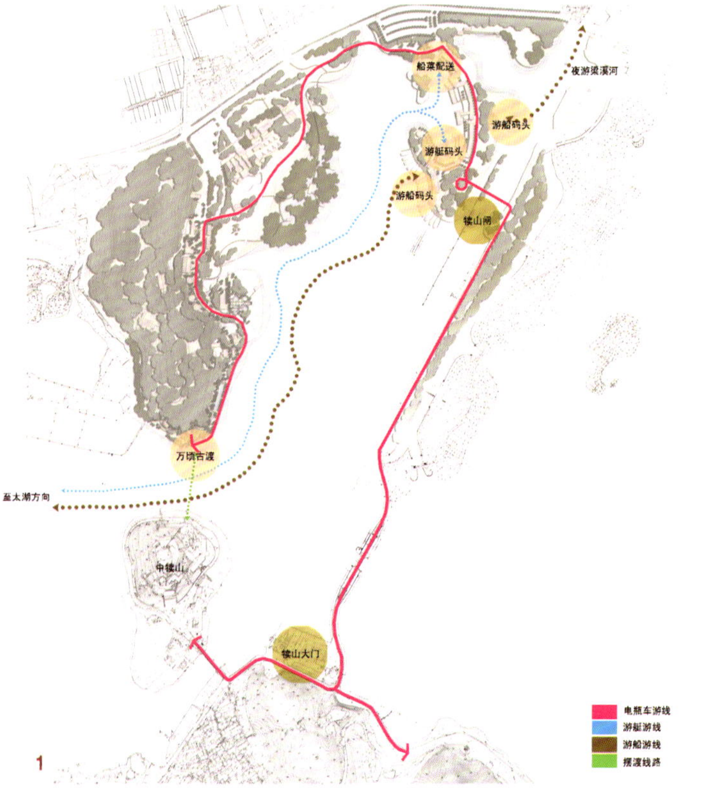
船菜配送
夜游梁溪河
游船码头
游艇码头
游船码头
犊山闸
万顷古渡
至太湖方向
中犊山
犊山大门
电瓶车游线
游艇游线
游船游线
摆渡线路

1

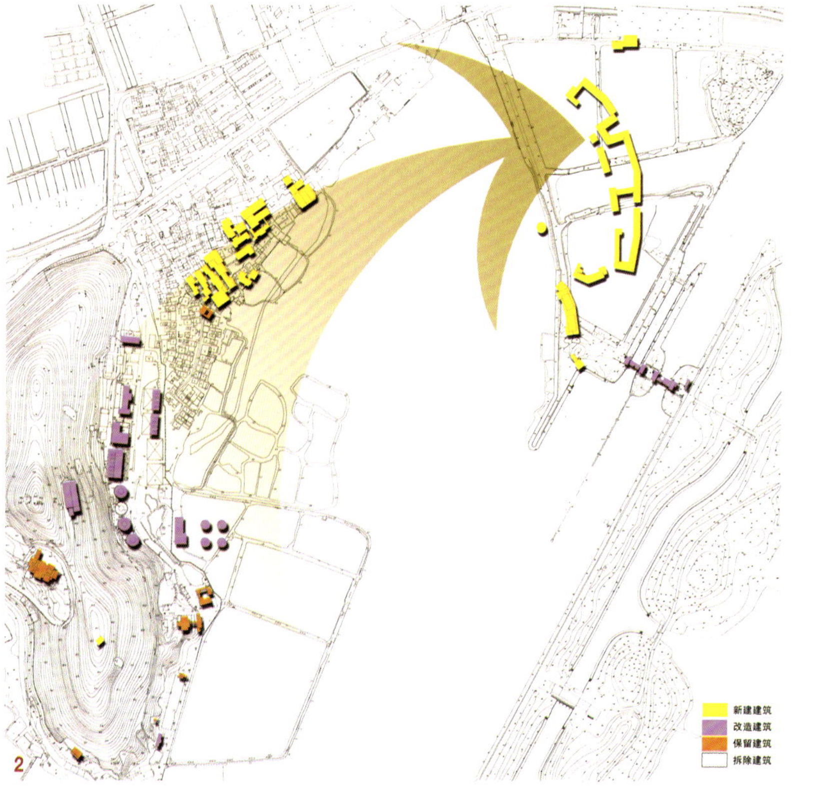
新建建筑
改造建筑
保留建筑
拆除建筑

2

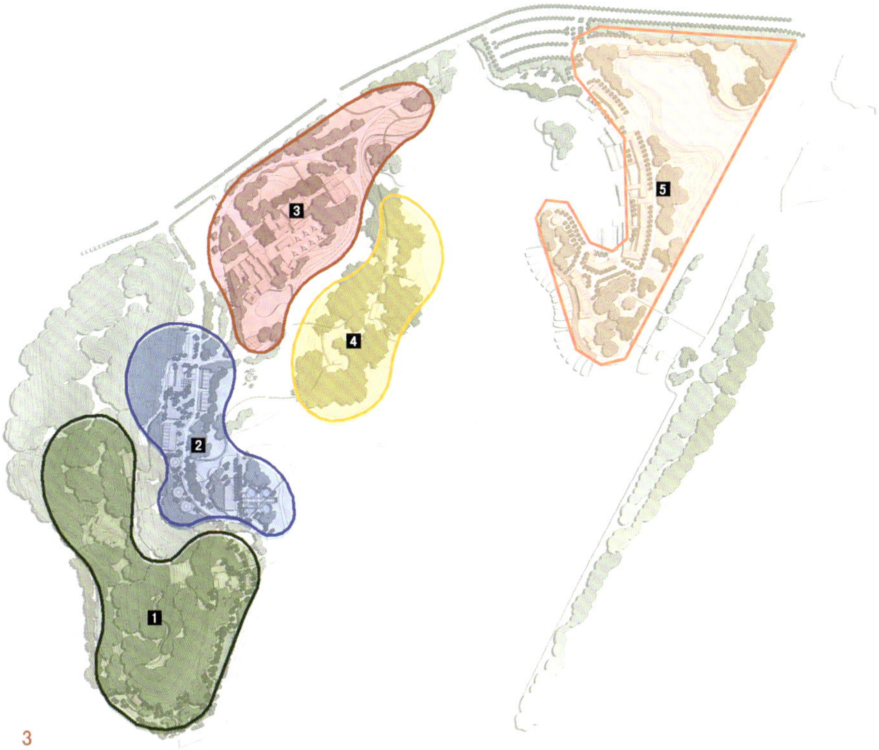

3

LAND INTERPRETATION

The lake is named after a famous figure who had a love story known to all Chinese. The key of the design is to explore the human connotation of the land.

PLANNING CONTENT

The design focuses on creating a fish village by integrating the resources of history, human and modernity. It involves the following aspects: to maintain the historic buildings; to add some new buildings; to transform partial buildings; to develop village complex; and to build some new sites for entertainment and sports.

DESIGN POINTS

History and culture section

The entire project tries to trim the tourism sites in a series order; to make the lake side road a first-class road, with some rest places being built in some flat areas. Yard and fence are introduced to trim the isolated buildings.

1. 交通分析图
2. 建筑时序分析图
3. 功能分区图
4. 鱻乡风情区鸟瞰图

4

Industry transforming section

The 4 clear water reservoirs are restored underground and only 4 clean water pond and a working path remain for visitors on the ground.

Lixiang flirtatious section

There are a lot of isolated yards in the Dongguan community, leaning against the hills and facing the lake. People there make a living by fishing. The design tries to rebuild the strategies, integrating the traditional and modern construction terms, and maintaining the existing harmonious relationship between the hills and river.

Water sports section

This section is isolated from others, reserved for activities like training, sports, entertainment and catering. The 2~3 storey buildings are built by glass, stone and steel, being open and compatible with the environment.

1. 渔人码头效果图
2. 游艇俱乐部效果图
3.4. “天地人合”工作室效果

1

3

4

工程档案

工程名称
无锡蠡湖管社山地区景观规划设计

工程地点
江苏省无锡市

设计单位
浙江大学建筑设计研究院风景园林分院

设计师
林畅雄、李瑛

项目景观总占地
40公顷

单位平米造价
220元

设计完成时间
2006年9月

竣工时间
在建

Landscape design of Wuyuan International Hotel, Shangrao city

上饶婺源国际大酒店景观设计

项目位于风景秀丽的全国旅游名胜区——婺源。这里人文文化、自然风景独特，可提供游客自驾游、自助游等多种形式的旅游方式。婺源国际大酒店要建设成为上饶地区高品质的酒店，必须具备高标准的软、硬件，因此，景观的配套设计显得尤为重要。

作为旅游地区的大酒店景观设计，不仅要有独特的自然资源，而且要有浓厚的历史人文资源，游客在此不仅可以放松心情，更能感受到独特的人文历史及当地的风土人情。景观设计要尊重当地的旅游资源，努力做到与周边风景相协调，尽量就地取材。

整个酒店坐落在一个小的山体上，东西落差为21m，酒店入口处低，具有较好的景观层次感。主入口主要以车辆进入为主，设计时加长入口通道的长度，使得车辆驶入酒店前，游客充满期待。主入口区域高差约为3m，采用了跌落式的停车场地的布置。

酒店客房前庭

酒店客房部的景观根据建设功能的布置和人流的方向，分为前庭、中庭和内庭院的布置方式，设计风格遵循徽派建筑的传统风格，景观亦走传统园林路线，给住客呈现小桥流水、曲径通幽的画面。在前庭景观中采用了酒店流行的大草坪设置，道路穿插其中，地形起伏变化。会议中心的分布包括停车位、主入口空间与人行通道，并且依山势设置了廊架。

在酒店规划范围内的最高点，设计了观景的特色山体景观，依山势造亭、修园路、设平台，形成了园路的起伏变化、空间穿插。在观景平台上可以观望对面的山体风景。在山凹中设置了高尔夫练习场地，起伏的大草坪与山峦相伴，彰显大气。在山顶布置了网球场，把现有的房子改造为活动服务用房。

在整个景观设计布局中，整体统一不乏变化，既遵循了传统中式园林景观文脉，又有现代气息的酒店文化元素。

1. 总平面
2. 高尔夫球场
3. 功能分区图
4. 酒店主入口通道

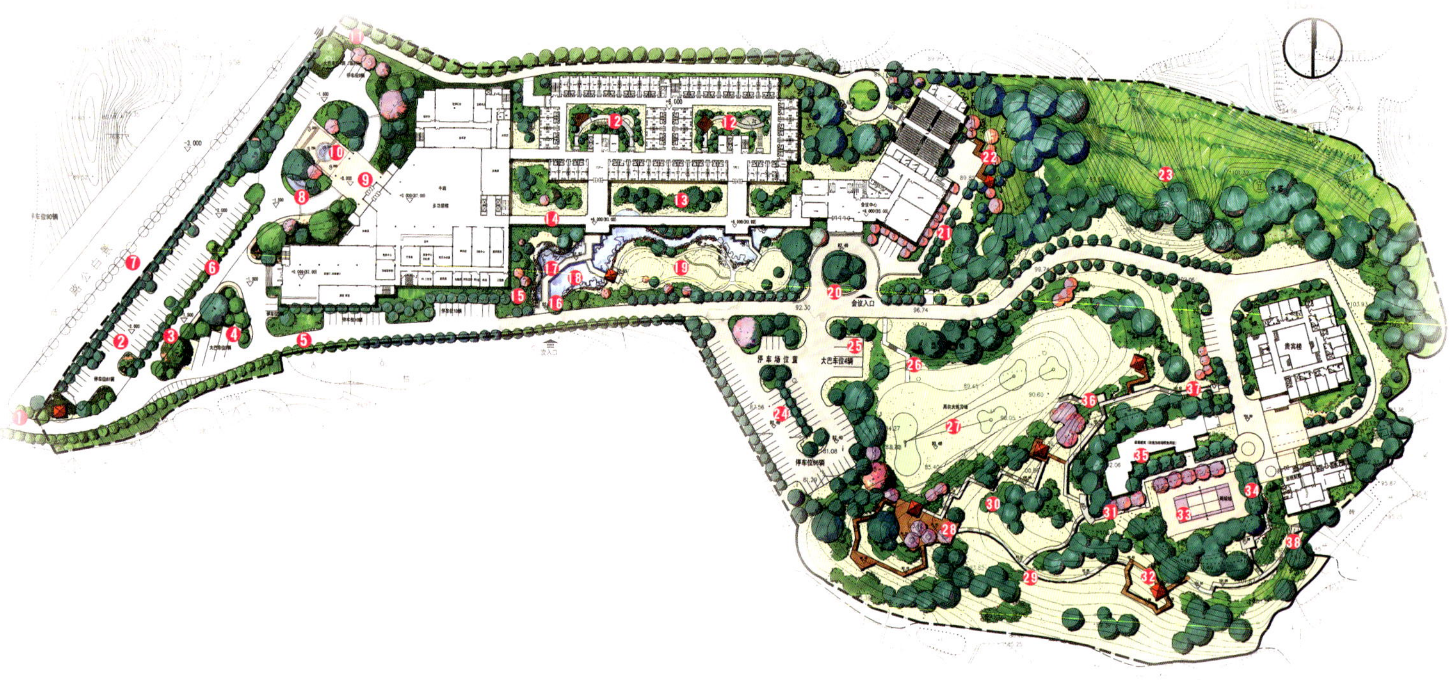

1

2

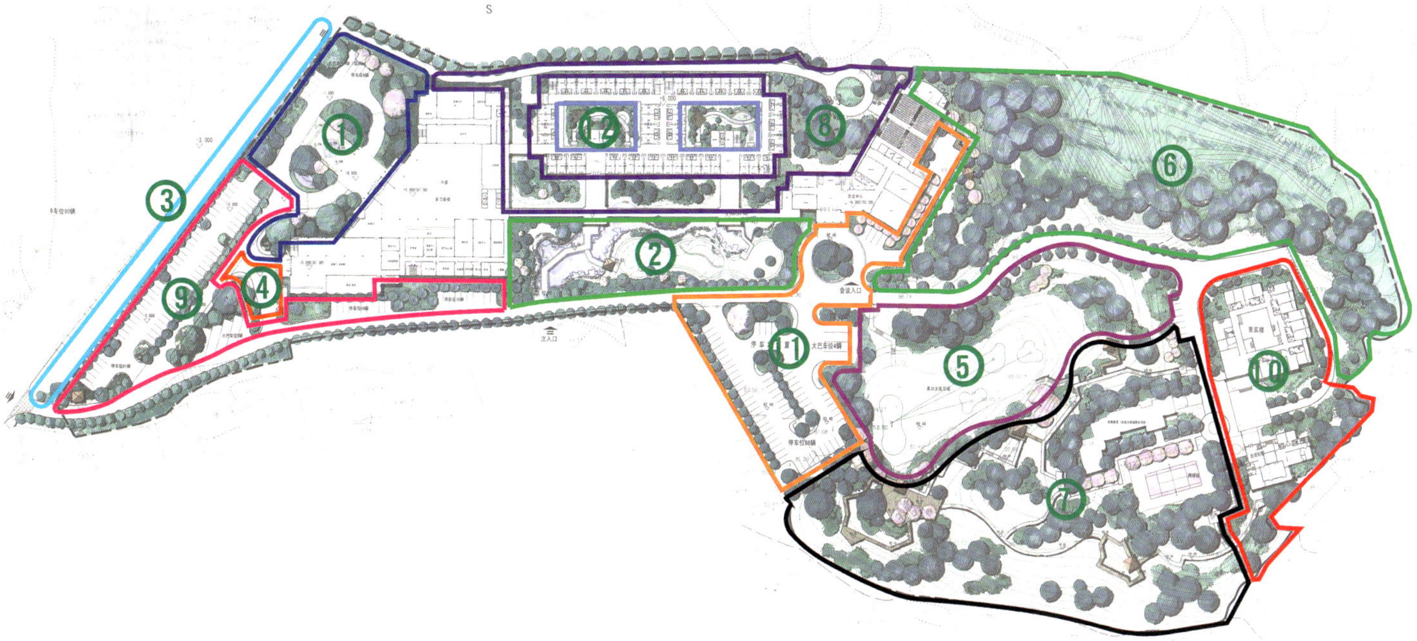

3

4

工程档案

工程名称
上饶婺源国际大酒店景观设计

工程地点
江西省婺源县

设计单位
杭州现代机构

设计师
丁志根、郭宏强、孟玮、徐中莹、洪燕飞、张敬涛、胡忠委、江小琴

项目景观总占地
6.96公顷

单位平米造价
150元

设计完成时间
2007年9月

竣工时间
2008年4月

The project is located in Wuyuan, a famous landscape area with its peculiar culture and nature characters where tourists can be served with various kinds of tourism tactics. To become a high-quality place in the region, Wuyuan International Hotel needs to have perfect scenery design.

The design needs to make people staying in the hotel feel the peculiar characters in the region.

The design highlights the eco nature. The hotel is located on a small hill, with a falling altitude of 21 meters from east to the west. The hotel entrance is lower in altitude level, providing better scenery vision. A long entrance corridor makes people sitting on the entering vehicles eager to see what is facing them. The parking sites in front of the main entrance are arranged in a falling manner where the altitude layer is 3 meters.

Inside the hotel are 3 courtyards which feature a kind of Anhui province-based manner. The scenery design adopts traditional approaches, including small bridge, water and curved paths. The front courtyard houses a big grass land with roads crossing on it. The designs of the conference centre includes parking sites, entrance area and sidewalk as well as corridor suspension

1

depending on the hill.

At the highest point of the designed area, there are scenery viewing kiosk, path and podium. One can watch the hill sceneries across the space. There is a golf training place in a cove. There is also a tennis court on top of the hill and the existing houses are transformed to serve the court.

The entire design contains traditional Chinese gardening touches and modern hotel elements as well.

1. 酒店主入口
2. 酒店次入口前庭

2

Landscape design of Lushanlian villa community, Changsha

长沙麓山恋别墅小区景观设计

小区车行主干道
小区车行次干道
景观步行道
城市主干道
道路主要节点

主要景观带
主要视点
主视点视线

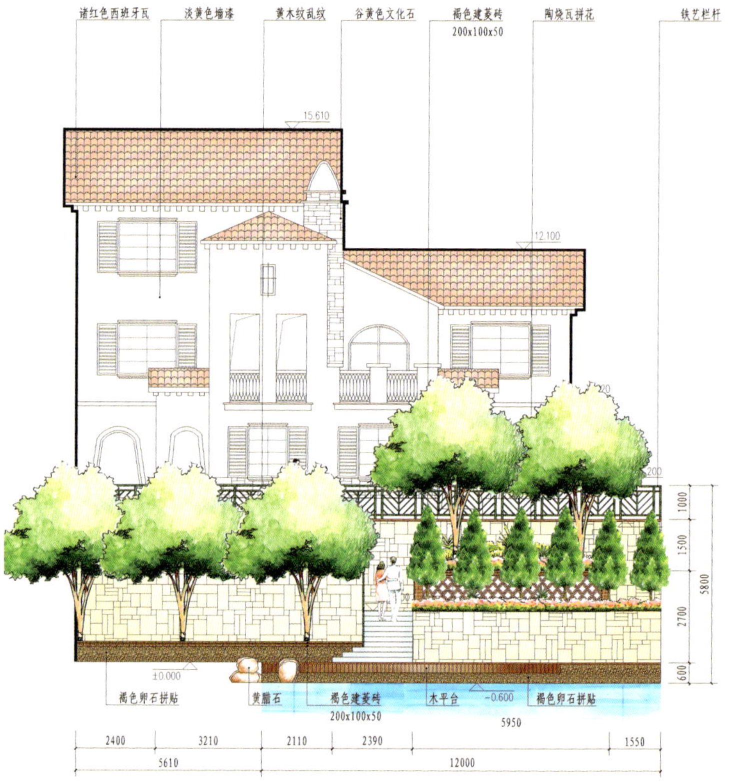
诸红色西班牙瓦
淡黄色墙漆
黄木纹乱纹
谷黄色文化石
褐色建菱砖
200x100x50
陶烧瓦拼花
铁艺栏杆
15.610
12.100
1000
1500
2700
600
5800
±0.000
褐色卵石拼贴
黄腊石
褐色建菱砖
200x100x50
木平台
-0.600
褐色卵石拼贴
5950
2400
3210
2110
2390
1550
5610
12000

台地亲水平台剖面图

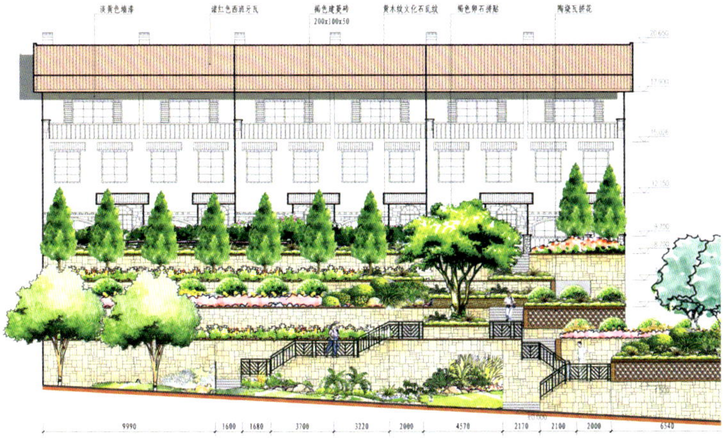

大阶级挡土墙立面图

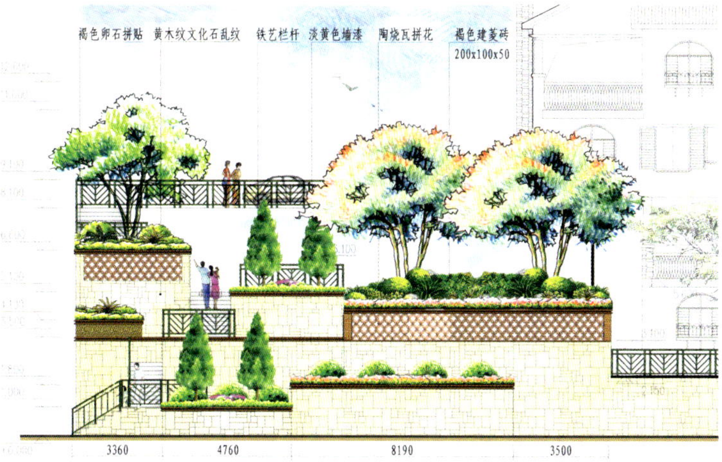

山顶挡土墙立面图

迪亚溪谷剖面图

麓山恋别墅小区位于长沙市望城县含浦镇，地处岳麓山风景名胜区，紧靠长沙市5000亩白鹤自然保护区，周围多为原生态自然林。项目距岳麓山科技园1km，距猴子石大桥约5公里，距省政府新地址约12公里。交通便捷，区域环境优越。

项目用地面积22公顷，有效规划用地面积20公顷，东西宽380m，南北长690m，地块较为规则，南向为香格里拉大道，西向靠三环线，东北两向为待开发的山地。场地内无建筑物，属丘岭山地，用地内有大小各异的四个湖面。用地东南侧是麓山别墅，其用地与本次规划用地东、南两侧相接，麓山别墅一期已建成，二期正在兴建中，区域形象较好。对本项目的开发建设起了一定的铺垫作用。

本项目的设计主题为“恋山·恋水·恋家”，体现纯正的西班牙风格。整个设计渗透着对自然朴素的西班牙山地庭院的眷恋、对幽静浪漫的卢佩特溪谷的眷恋、对朴实高雅的异国艺术情调的眷恋。

南大门效果图

1

2

3

1. 野外平台
2. 售楼部后景
3. 别墅花园
4. 卢佩特溪谷透视图
5. 山路端点小景

4

整个项目总体布局由一核心、三带、四大异国风情组团，若干景点组成。一核心指"迪亚湾"，用地中心湖面，是以会所、休闲设施构成的小区公共中心；三带指靠近三环线的隔噪声绿化景观带、环形樟树和银杏景观大道（园内两条主干道）；四大异国风情组团分别为：贝萨卢组团（西班牙文化艺术景观区）、索列尔组团（西班牙山林庭院景观区）、卢佩特组团（卢佩特浪漫溪谷景观区）、梦塔那组团（迪亚湾景观区）。

在具有千年山水灵气的长沙岳麓山下，一座纯正西班牙风格的小城——麓山恋·迪亚溪谷，让远离喧嚣的人们能享受到天堂般的美景：让您生活于自然朴素的西班牙趣味庭院中……让您漫步于幽静浪漫的卢佩特溪谷中……让您陶醉于朴实高雅的异国风情艺术中……

5

1

2

1. 交流凉亭立面图
2. 山顶广场
3. 高迪文化休息小空间透视图
4. 森林休息花架

3

工程档案

工程名称
长沙麓山恋别墅小区景观设计
工程地点
湖南省长沙市
设计单位
长沙筑园建筑景观设计咨询有限公司
设计师
王小保、古德泉、简天佐、许海南、范米娜
项目景观总占地
22公顷
单位平米造价
320元
设计完成时间
2006年12月
竣工时间
2007年12月

Lushanlian villa community is located in Hanpu town, Wangcheng county of Changsha. Inside the Yuelushan scenery area, it is very close to the 5000 mu white crane protection area of Changsha, surrounded by forest. The project is 1 km away from the Yuelushan Sci-tech Park, 5 km away from the Shizishi bridge and 12 km away from the newly located provincial government. It has easy traffic assess to many places.

The project plans to use a 22 hectares land, or 380m X 690 m. The land is surrounded by the Sangrila avenue in the south, the 3rd ring road in the west and undeveloped lands in the east and north ends. There was no buildings on the land and 4 pieces of lake. In the south-east end of the land is lushan villa project. The first-phase buildings have completed constructions and the second-phase constructions are being done. The sound appearance helps upgrade the very project.

The theme of the project is designed to represent pure Spanish characters. The entire design is sentimental attaches to the Spanish compounds, the river and the foreign country

Other designs include a centre, referring to the central lake which supports a resort centre and leisure facilities; the three belts refers to the three plants belts close to the 3rd ring road and alongside two main avenues inside the park. There are also 4 foreign touch complexes.

4

Landscape design of Yizhao Xiayang Yinhu villa community, Suqian city

宿迁亿兆夏阳·银湖别墅景观设计

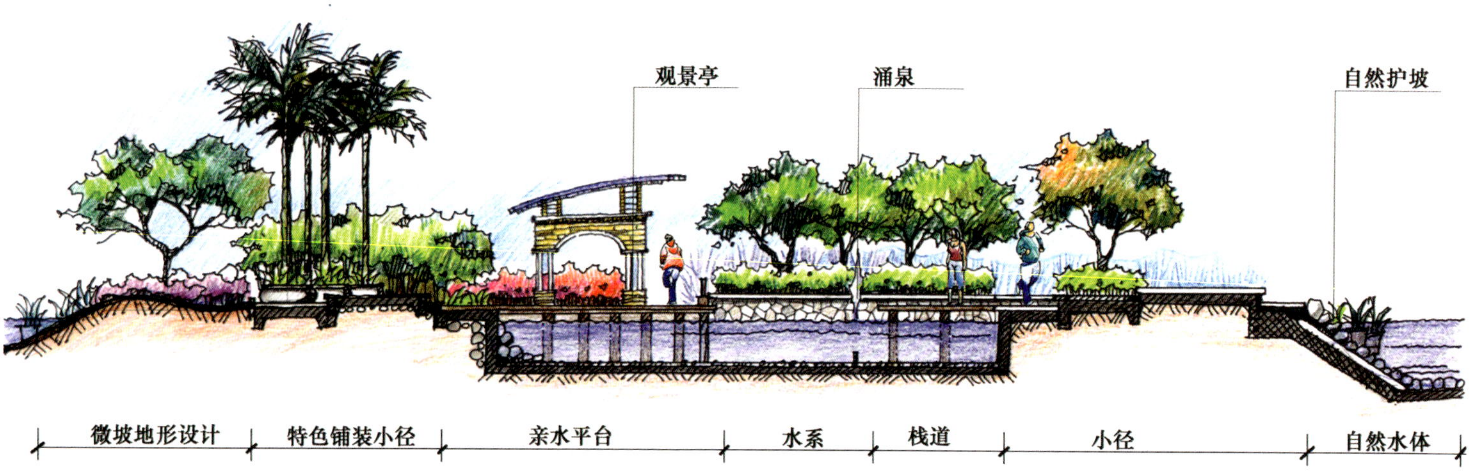

1

主题的确立

项目设计围绕现代北美别墅风情园这一主题展开，以北美邓肯小镇、维多利亚港湾、加拿大红枫谷、科罗拉多风情园为主标题，同时配有蒙特利尔花园、温哥华浪漫城、魁北克异陆"枫"情、梅尔维尔半岛、格兰佛岛、纳帕溪谷、维多利亚沙滩、梅萨绿谷、奥罗拉庄园等组团区域，生动的再现异国风情园。将东方的人文思想与西方的居住风格有机结合，使得"绿脉""蓝脉"和"文脉""山脉"相互渗透，集居住、度假、休闲、康体于一体，向往自然、注重健康，追求自由、私密的领地，体现健康、快乐、生态的现代生活理念

概念设计

我们在西方园林的个性与秩序、东方园林的"师法自然"中找到了契合点，并把它放入本案的景观设计中，倡导建筑与景观园林的自然延续，追求坡地别墅的高低起伏和层次感，将绿色引入家中，超大面积的外庭院与内庭空间互映，加强别墅与花园的渗透性，园中有别墅，别墅中有园，营造平和宁静、视野宽广的北美小镇情调，处处弥漫着舒适浪漫的生活气息。

四周绿荫环抱，遗世独立的私密性别墅使住户隐没在一片绿色之中，户与户之间绿化充斥、大树环绕，私密性绝佳。在每栋别墅设计过程中考虑的不仅是每个小庭院自身享用，更是设计出院与园之间，院与园之间的互相借赏，外围密闭中间开阔，园内充满北美西部原野的神秘。茂密丛林、山泉迭水、旷野台地，地形层层迭起、气势雄伟壮观，将人造景观与周围的环境有机结合起来。

总体布局框架

结合地形、资源特征，对东高西低的外围环境，采用湖、港、岸、岛、水溪、山地六级递进式的空间布局，分级处理，开放与私密兼备，风景与人文并存，有序与自然对话，用点、线、面的手法有机的联系起来，以骆马湖为伴建一个具有维多利亚港湾风情的集度假、娱乐、旅游为一体的"夏阳银湖湾"，以嶂山为背景建一个具有时代生活气息，追求文明、健康、快乐生活方式的别墅园。

2

1. 维多利亚港湾景观节点剖面图
2.3. 梅萨绿谷岛屿临水别墅

3

1

2

工程档案

工程名称
宿迁亿兆夏阳·银湖别墅景观设计

工程地点
江苏省宿迁市

设计师
金晶（南京艺术学院）

项目景观总占地
22公顷

单位平米造价
120元

设计完成时间
2006年12月

竣工时间
在建

1. 景观总图
2. 格兰佛岛景观节点
3. 维多利亚港湾临水别墅
4. 魁北克异陆“枫”情
5. 梅萨绿谷岛屿临水别墅

3

4

5

1

ESTABLISHMENT OF THE THEME

The project design focuses on touches in north America, combining the eastern ideas and western living style, making green, blue, human and mountain factors penetrative to each others. The villas are design to integrate living, vocation, leisure activities and sports.

CONCEPT DESIGN

The designers find a connection point among western garden and eastern garden. They advocate the natural sustaining of buildings and the garden, introducing green into houses, making the outer and inner yards compatible. The result is that was villa in garden and villas house garden.

In the design of each villa, the considerations fully cover the mutual benefit of yard and garden and dynamically integrate the artificial scenery and the environment.

ENTIRE LAYOUT STRUCTURE

Based on the landform and resources, a 6-steps space arrangement is introduced, including lake, port, bank, island, brook and hill and the point, line and face are connected dynamically. The Xiayang Yinhu is based on the Luoma Lake, houses vocation, entertainment and tourism activities. The villa area takes the Zhangshan as a background, creating an environment of civilized, healthy and happy.

2

1. 坡地水岸
2. 科罗拉多之舟
3. 入口
4. 魁北克异陆“枫”情
5. 梅萨绿谷中心景观
6. 奥罗拉庄园
7. 梅萨绿谷中心景区

3

4

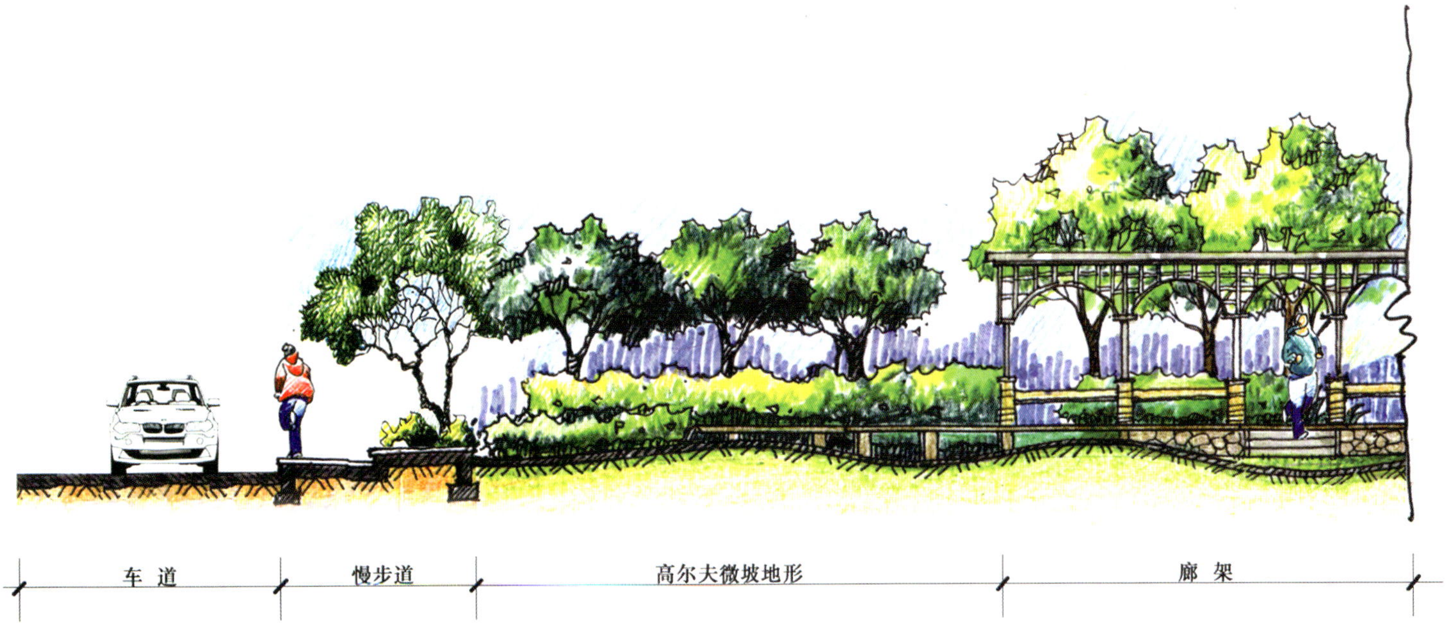

5

6

Landscape design of Zhongshanmeilu villa area, Nanjing

南京钟山美庐别墅区景观设计

仿浅滩自然景观

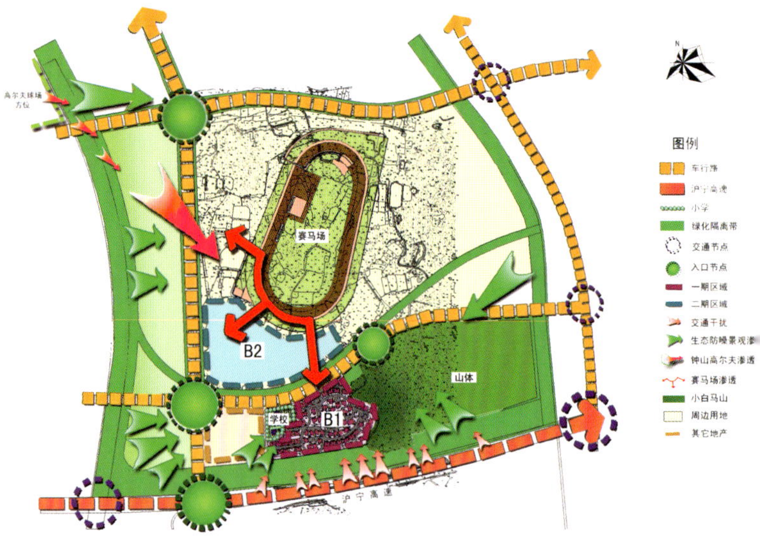

项目位于南京市东郊紫金山东麓的国际赛马场旁边，它东依白龙山，南望沪宁高速公路，西眺紫金山，北望南京国际高尔夫球场。

项目定位

项目充分利用基地周边的独特资源和深厚文脉，旨在创建一个主题突出、体现高尚运动与时尚休闲完美结合的健康型理想居住场所。设计中着重体现山居、庭院、民国品质与文化精髓，将山水文化渗透到现代都市生活中。

景观规划分析

按照大景观原则，设计师从项目的总体规划布局入手，将社区的建筑和景观以周边自然丘陵坡地环境为依托，由东至西，自北向南保留原有峰峦山脚的自然起伏，以多层次变化的建筑和景观植物的分隔，构筑出山庄淳朴清逸的独有气质和山林坡地式社区景观风格。

公共空间

山庄内水景观及建筑标高结合场地内的地势、地貌、高差营造出仿溪谷、叠水、涌泉、池塘、浅滩等自然景观，并构成了联系贯穿全社区主要景点的双"V"骨架及中心景观轴。

社区内主要道路景观采用微地形、护坡挡土墙及封闭式绿化相结合的形式，利用弯曲的道路，起伏的地形和郁闭围合及开畅流通的变化，营造出绿色环绕、幽静深邃、典雅别致的景观氛围。

会所公共景观及主入口景观是社区对外联系与整体印象体现的重要区域，在场地环境有限的情况下，紧密结合会所建筑形态与功能，通过水景、铺装、景墙、花

草树木与小品，创建简约时尚、精巧细致、功能完善、展现民国主题风格和人性化设计的景观环境。

私家区域

别墅小院及私家花园景观充分体现私家生活空间和邻里交流空间的有机结合，体现个性化、人性化、舒适性和时尚性。通过对功能化景观的精心设计，体现与建筑分享品质、与自然分享健康、与家庭分享亲情、与庭院分享生活的设计理念。

景观小品

公共景观和社区内部景观中的景观构筑物与小品丰富了社区环境的文化内涵，提升了景观品质。小品设计以反映民国主题与体现时尚休闲文化为内容，利用柱灯、石雕、石墩、花台、水钵、景墙及私家庭园等，自然有续而且富于变化

1. 总平面
2. 一山一水，一木一径，对一代经典无声承袭
3. 景观营造私密空间
4. 承载最醇厚的民国建筑精粹
5. 大树护守屋檐，青藤爬上路上

的分布在社区不同的景观区域。

植物景观

社区植物设计考虑大环境景观因素并遵循适地适树原则，分层次，分重点，由内到外，植物品种的选择从细致到粗犷，从丰富变化到协调统一，由私家花园的庭荫树到社区组团的主景树，从社区内部的基调树种到环绕周边的植物组群都尽量考虑地域特征与景观特色的关系。

高速绿化带

高速绿化隔离带是以周边丘陵坡地环境为背景，利用高速路隔离带用地人工堆造地形和多层次绿化种植，形成对社区场地的自然形态的围合，起到屏蔽不良视线及消弱声波干扰的作用，对外突出展现整体建筑形象和景观效果，同时，也为社区居住者提供了一个茶余饭后休闲散步的景观场所。

1. 景观营造私密空间
2.3. 私家庭院隐于浓荫下
4. 半私密空间
5. 浓荫守望的百年国府

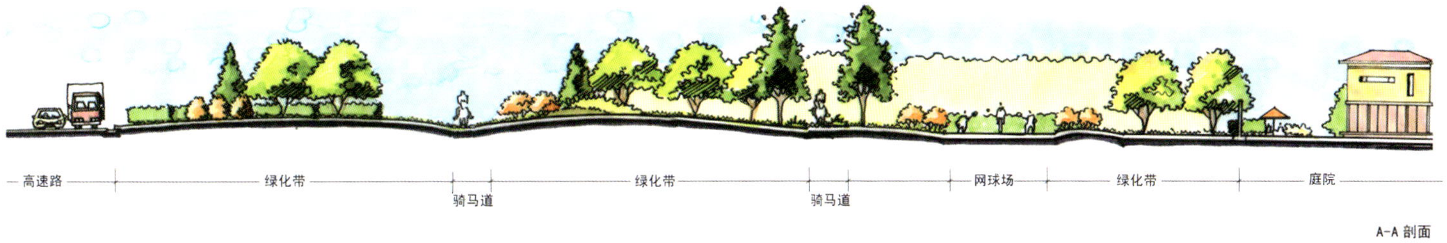

The project is located beside a race track of the east end of eastern Nanjing, with Bailong hill, Nanjing-Shanghai expressway, Zijin mountain and a golf course around it in its four ends.

PROJECT POSITIONING

The project makes full use of the peculiar resources and traditional cultural heritage around to create a residential community highlighting the theme of fashionable sports and leisure activities. It focuses on the mountain residence, compounds, cultural soul, penetrating the traditional Chinese culture into modern life.

AN ANALYSIS ON THE SCENERY PLANNING

The designers managed to maintain the natural waves of the mountains from east to the west and from north to the south, as well as the isolation of the leveled buildings and scenery plants. The villas are imposed with the simple and leisure temperaments and typical characters in the mountainous areas.

Public space

The dale, stacking water, boiling spring, pond and riffle inside the project is situated to the topography and landform, forming a double V-shape frameworks and a central scenery axle.

The road sceneries adopt the forms of revetment, retaining wall and enclosed greening area. Curving

5

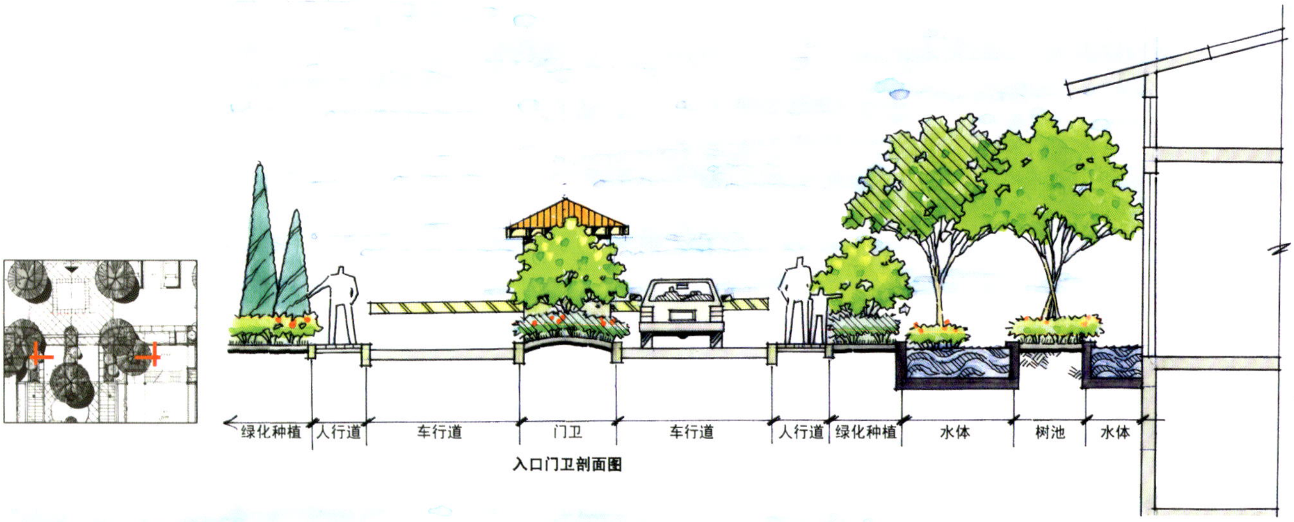

入口门卫剖面图

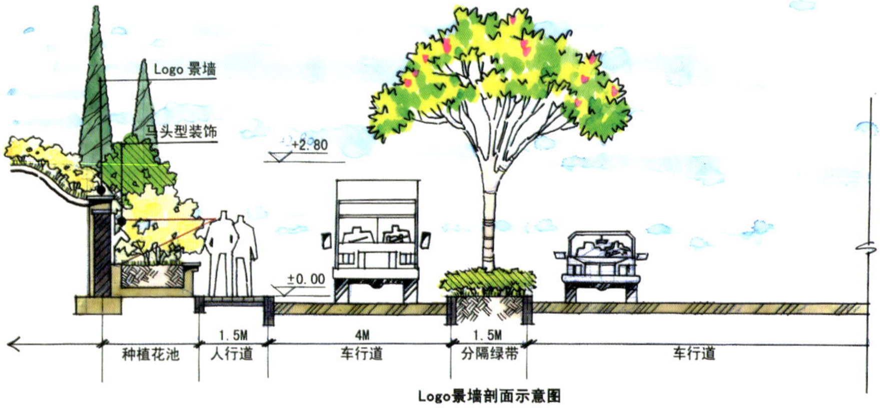

Logo景墙剖面示意图

工程档案

工程名称
南京钟山美庐别墅区景观设计

工程地点
江苏省南京市

设计单位
易兰亚洲规划设计事务所

设计师
陈跃中

项目景观总占地
6.08公顷

设计完成时间
2007年1月

竣工时间
2007年9月

way, up-and-down landform and other changes help create a scenery environment that is green, quiet and elegant.

The chamber and entrance sceneries are important for the socializing activities and displaying its image. Water ornament, pavement, scenery wall, grass and tree and ornament are introduced, within limited area, to display an environment that is fashionable, precise and perfect in functions.

Private area

The courtyard and garden fully display a combination of the family life and neighborhood friendship.

Scenery ornament

They refer to column lighting, stone sculpture, stone stool, flower platform, water bowl, scenery wall and gardens. They are changeable and are placed inside the area in an orderly manner.

Plant scenery

The choice fully considers the entire environment and very place where the project is located, with regard to the private park and the community, inside the community and the area outside it.

Expressway greening belt

The greening belt is designed to screen the disorder vision and noise. It is also a good place for the residence in their spare time.

入口区景观

景观营造私密空间

Landscape design of Guantang residential community (second phase), Beijing

北京观唐别墅区二期景观规划设计

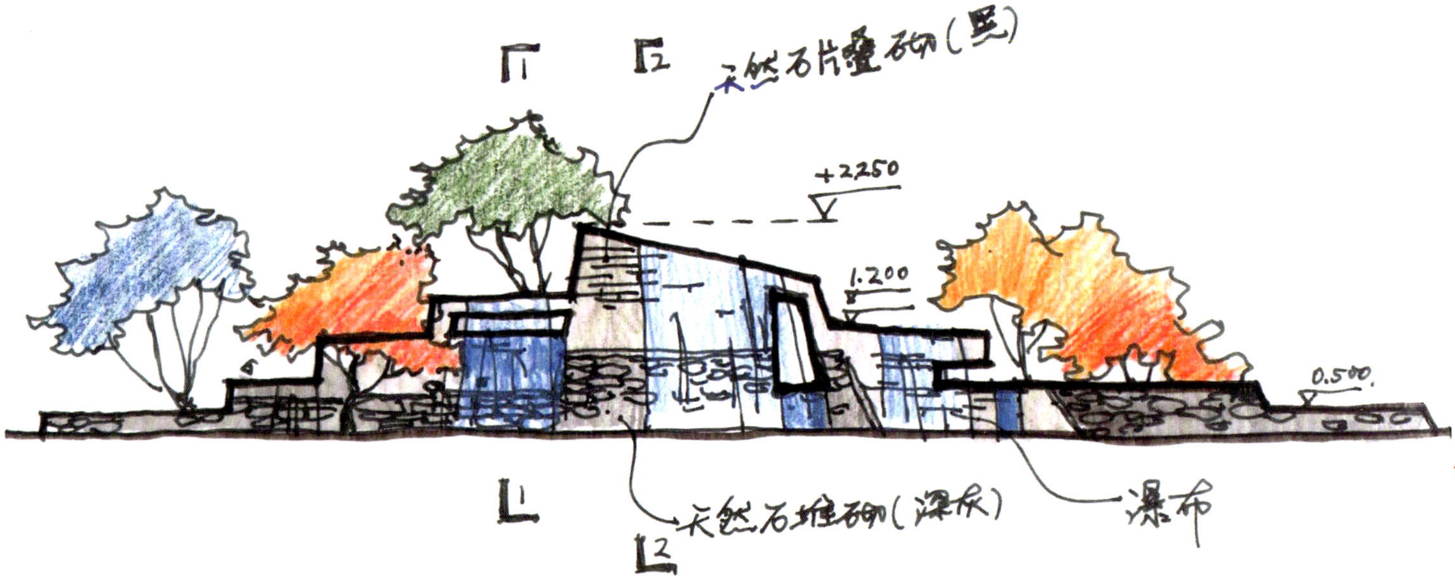

1

观唐别墅区位于北京市朝阳区东郊农场中心区，以中式庭院式建筑布局为其主要特色。已建成的一期建筑、景观、室内均为古典的中式风格，二期的建筑方案也延续一期的建筑风格，景观规划的定位为新中式景观风格。

通过对现代别墅生活与古典园林文化的分析选择设计切入点：古典园林文化气质与观唐别墅生活的结合；在喧闹拥挤的现代生活中对古典意境的追求；对古典造园手法的转换与运用。

设计师在对整个园林景观分析定位后，对各个分散的庭院绿地进行了主题定位，这也是古典园林突出主题节点的设计手法，并以街道景观将各个分散的绿地联系起来，形成完整的景观序列。每个庭园通过对地理位置、建筑环境、空间场地特性的分析形成自己的主题。分列如下：

中心步行街

是整个观唐别墅区的中心，其意境定位为"文城街影"，营造"街"的繁华与氛围，成为整个别墅区文化交流的中心场地。

府后花园

鉴于场地特征，方案定位为静谧、精致的私家园林，为东西附近住户的私家庭园。

南侧绿化带

具有开阔大气的皇家园林文化气质，注重景观建筑的空间视觉关系与环境的文化生活承载需求。

英式庭园与法式庭园

营造异域文化特征，唤起有海外背景住户的文化认同感。

西侧绿化带

原始山林意境，营造生态的山林环境，吸引各种生物在此繁衍生息。

涵碧春霏

以春天的烂漫为主题，满园的红玉兰将在春天绽放，表现纯粹的桃源意境。

香洲夏雾

以夏天的璀璨与欣欣向荣为主题，池中荷花飘香将成为庭园的主要亮点。

街道绿化

努力营造老北京胡同的景观空间效果与生活氛围，各个住户门前都有自己的特色种植景观。

2

1. 入口标志景观手稿
2. 彩雾园
3. 府后花园

3

1

2

3

4

5

6

1. 总平面
2. 中心步行街
3. 涵碧春霏
4. 南侧绿化带景观——木平台
5. 南侧绿化带景观——滨水长廊
6. 南侧绿化带景观——四角亭
7. 涵碧春霏平面图
8. 彩雾平面图
9. 法式庭院平面图

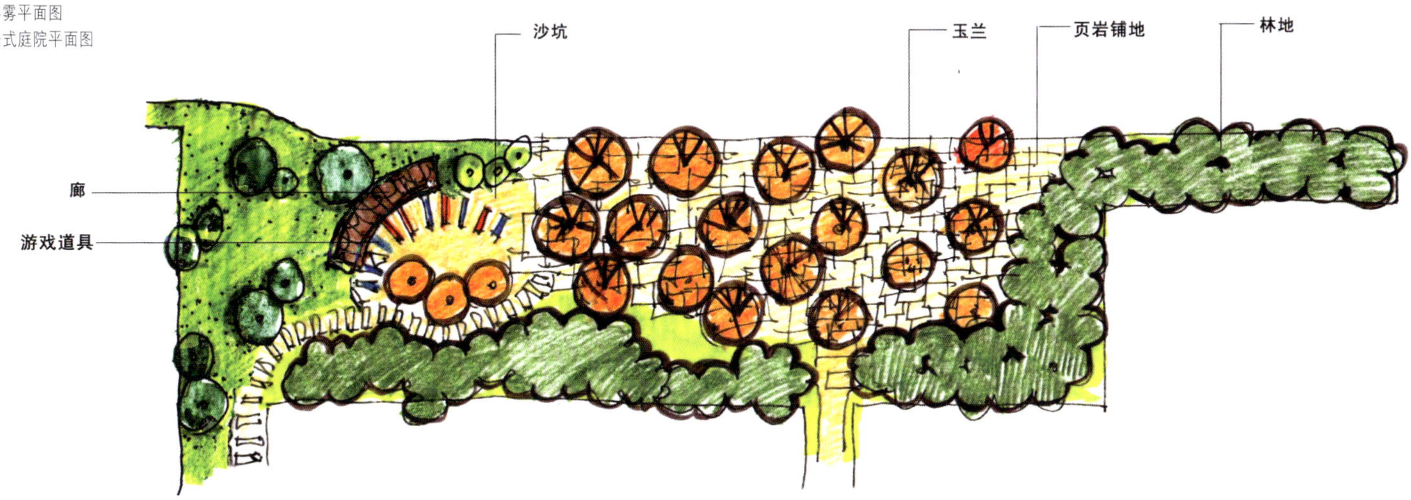

7

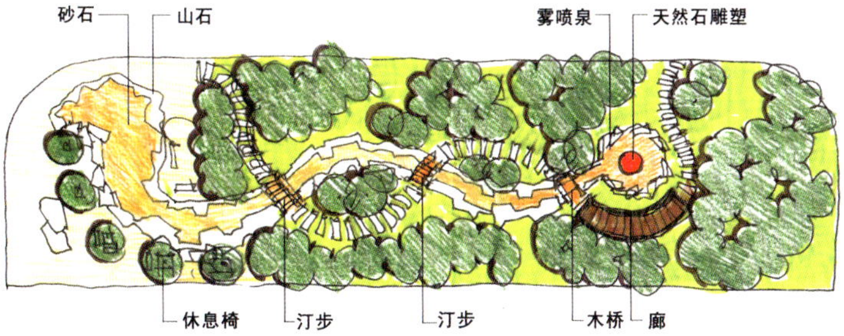

8

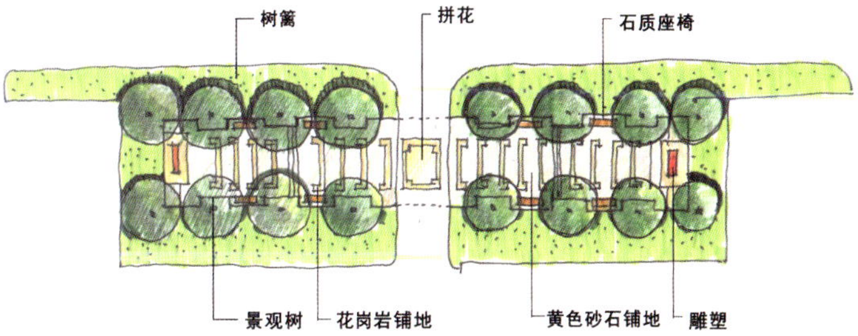

9

1

2

The project is located in the central Dongjiao farm in Chaoyang district, Beijing. The first-phase buildings, sceneries and interiors are typical Chinese style and the second-phase project is expected to continue the positioning.

The point is: to pursue the traditional things in modern life and the methods will be a conversion and application of the classical gardening way.

The designers manage to position each isolated yards and the street sceneries are used to connect the scattered grasslands and each yard has its own theme.

Central pedestrian street

It is the centre of the project for cultural exchanges.

Backyard garden

A quiet and concise garden for residents around it.

South end greening belt

A typical royal garden, focusing on the spatial vision relationship of the buildings and the environment.

British and French style garden

Appealing to those moved from outside China.

West end greening belt

The original forest is expected to accommodate all kinds of wildlife.

Yulan garden

A typical peach garden full of red Yulan flowers.

Summer garden

A typical water lily garden

Street greening

A typical street of an ancient Beijing where each house is trimmed with peculiar plants.

3

4

5

1.2. 观唐建筑风格
3. 入口标志景观
4. 香洲夏雾
5. 南侧绿化带景观——入口拱桥
6. 中心步行街
7. 项目区位图
8. 泳池铺装

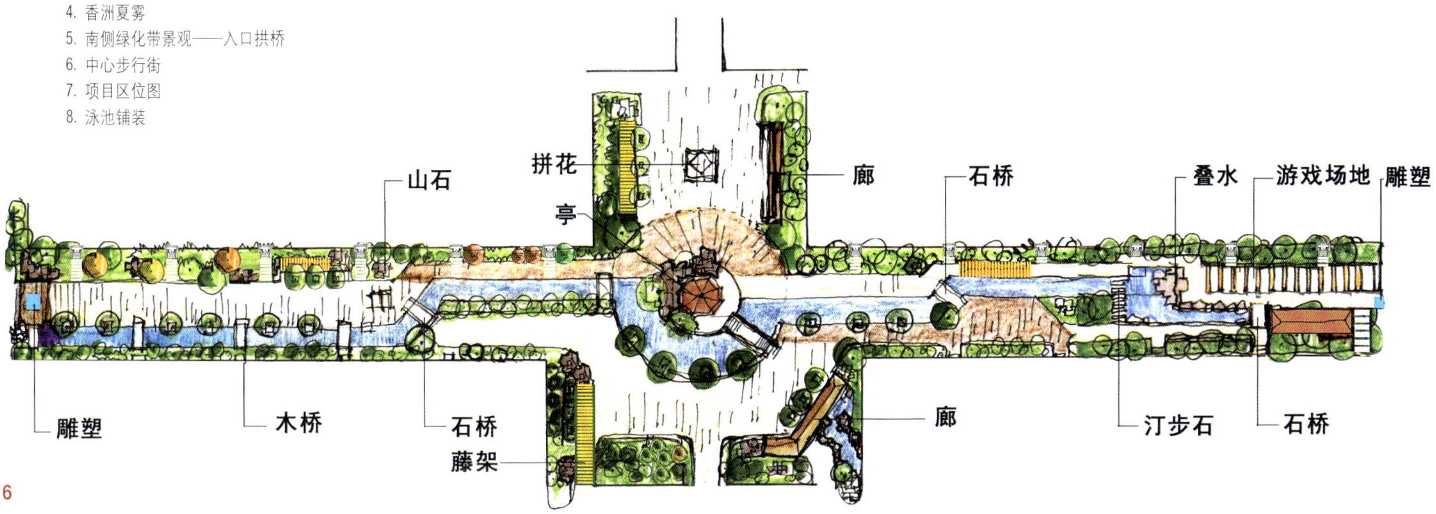

6

7

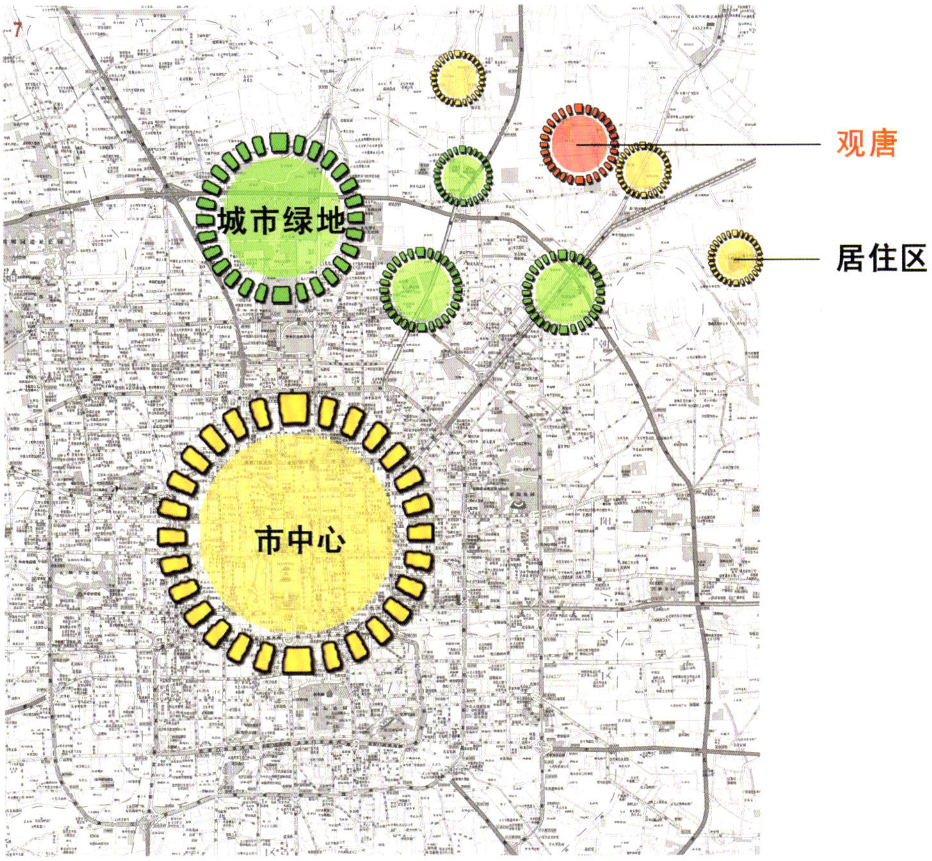

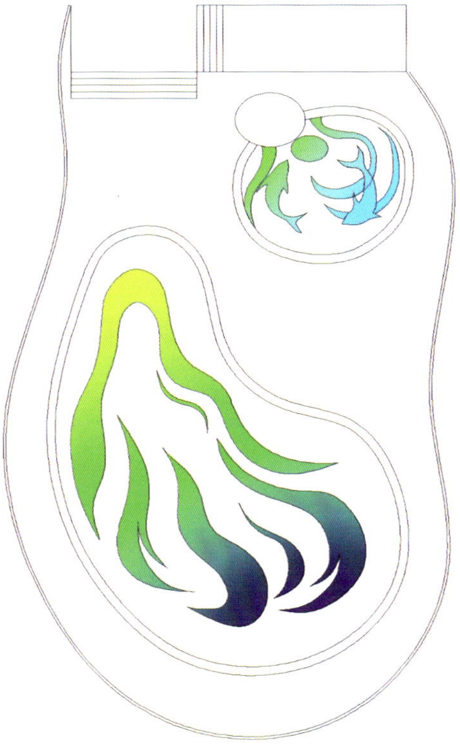

8

1

工程档案

工程名称
北京观唐别墅区二期景观规划设计

工程地点
北京市朝阳区

设计单位
北京北林科贸实业总公司

设计师
于立晗、柴冠杰、刘春华

项目景观总占地
4公顷

单位平米造价
500元

设计完成时间
2006年7月

竣工时间
2008年3月

1. 中心步行街
2. 南侧绿化带景观——瀑布
3. 香洲夏雾平面图
4. 府后花园

2

3

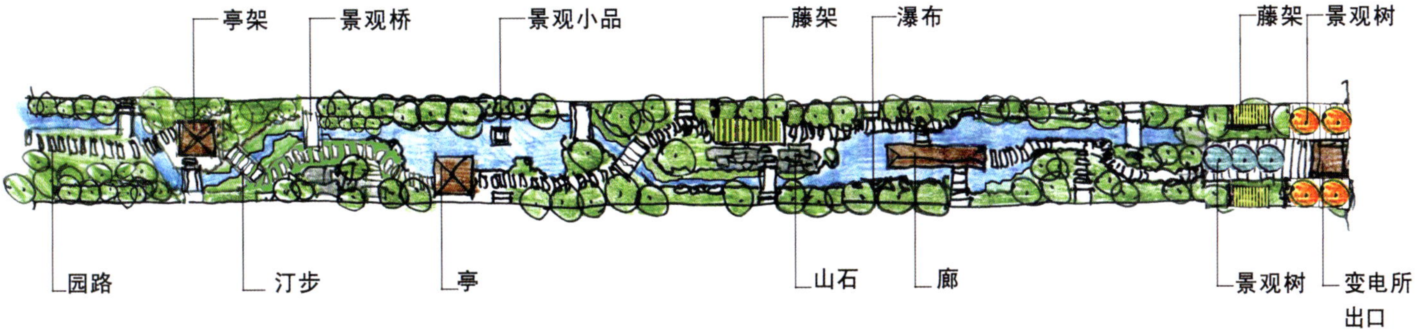

4

Landscape design of Yishuzhuangyuan, Xiangtan city

湘潭逸墅庄园景观设计

1

项目概况

湘潭逸墅庄园地处长潭双城轴心门户位置湘潭河西九华工业区，项目占地约30.9公顷，东西跨度370m～540m，南北跨度680m～730m，东临富洲路，西接宝马路，西向和北向分别为城市规划道路，距离省会长沙26km，仅18分钟车程，是"长－株－潭"城市群区域规划、湘江生态经济带规划的重要战略发展区域。该地段生态平衡、空气清新、山形秀美、植被丰富，自然环境优美，交通四通八达。

场地分析

南北15m高差、东西10m的高差自然地形以及两个面积约8000m²的水体和舒缓的山峦沟谷是本案基地现状的重要特征。逸墅庄园的总体规划凭借其得天独厚的自然条件，在方案规划中最大程度保留了原生态的天然水体、地形地貌以及自然植被，以充分发挥大尺度的真山真水和台地缓坡的有利条件，目标是将逸墅庄园打造成顶级的庄园别墅景观社区，让这里的原生山林、坡地、内湖之上的七大别墅组团、两大宫廷式城堡主题会所、两大印象派艺术主题湖泊水系景观及艺术风情主题园林结合尖塔、圆拱、列柱、喷泉、广场、雕塑、庭院等景观建筑，散发出古老欧洲皇家贵族的神韵。

根据现场调查、对一期以及总体规划的研究，我们认为本项目具有以下一些有利

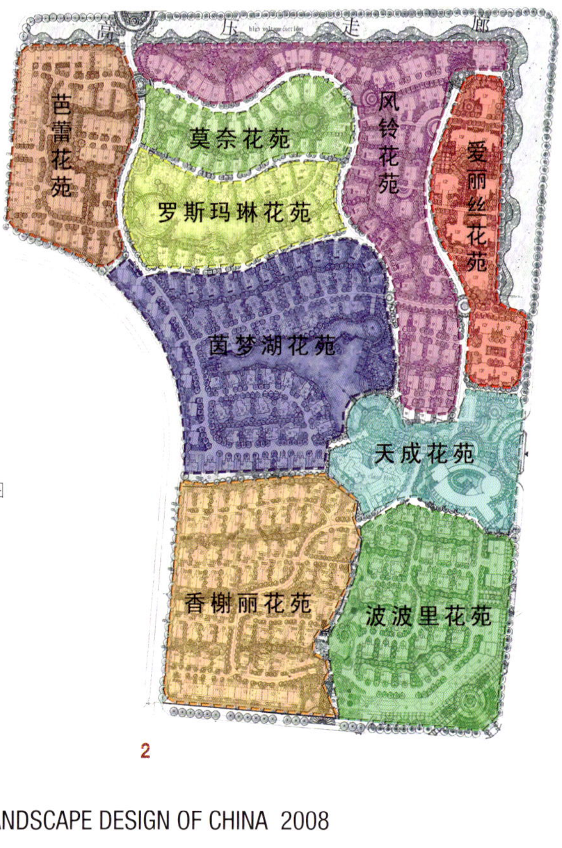

2

1. 总平面
2. 景观(组团)分析图
3. 景观结构(街道)分析图
4. 主入口逸墅广场
5. 茵梦湖花园鸟瞰图

3

4

5

因素和限制条件。

有利因素

别墅区绿地组团因地制宜、布局合理，能够做到疏密有致，有利于花苑组团内园林景观的空间组织；小高层组团位于总体布局的东北向，对别墅庄园的整体风格少有影响，其内部布局也较为合理，为别墅组团景观提供了较大的延伸渗透空间。

自然天成的两个湖面面积较大，水质条件较好，为湖边顶级别墅以及花苑组团景观提供了优越的地理条件，不同高差和不同形状的两个湖面有利于湖面水系景观的差异化和系统化组织。

伴有较大湖面的缓坡别墅台地空间为营造古典欧洲庄园景观提供了有利条件。

限制条件

茵梦湖的常水位标高与南北两厢别墅正负零标高相差1.5~2.8m，罗斯玛琳的常水位标高与南北两厢别墅正负零标高相差4.7~7.7m，如此大的高差为湖滨别墅营造一定私密性的同时给亲水景观带来了较大的难度。

庄园会所前广场北向道路有一段40m 17%左右的坡道，给园林景观和消防带来

1

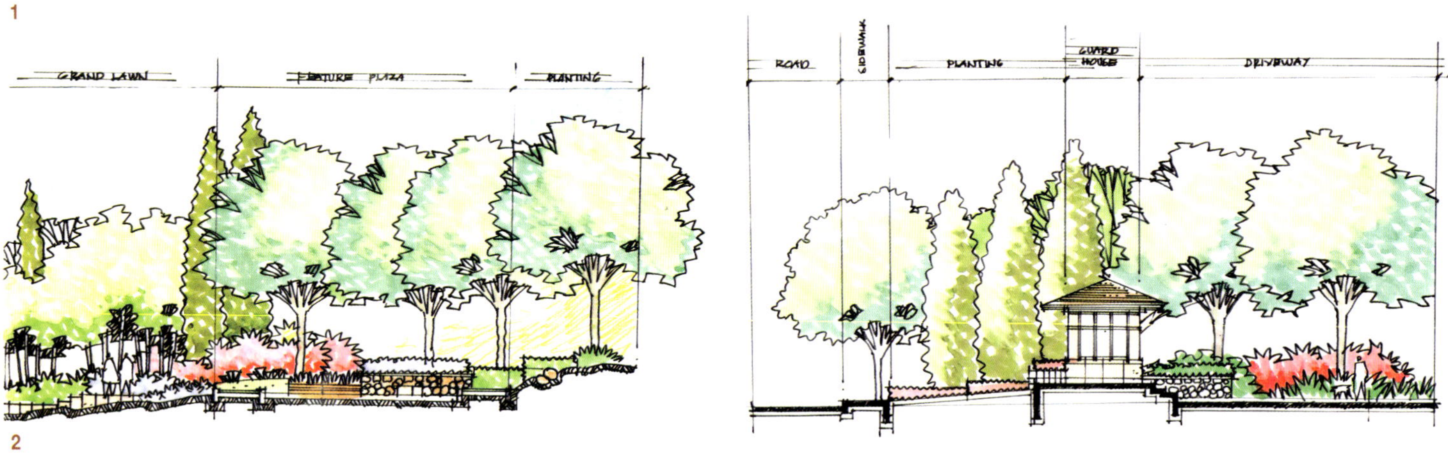

2

一定的难度。

部分别墅组团绿地过少，不利于功能性场所的安排。

设计定位

依据“逸墅庄园”的项目定位，其建筑风格偏重于北美风情，其景观园林风格也以欧美风情为主。因此，“逸墅庄园”的设计定位必须从项目总体的庄园风格与艺术氛围的提升出发，合理分区，建立分区的景观理念，丰富景观整体风格下的多样性，营造景观的均好性。

1. 贵商业广场立面图
2. 日出花园中心区剖面图
3. 天鹅堡景区
4. 圣地街鸟瞰图
5. 香榭丽大街
6. 拾梦桥景区
7. 伊丽莎白街
8. 湖畔花园

3

4

5

6

8

1

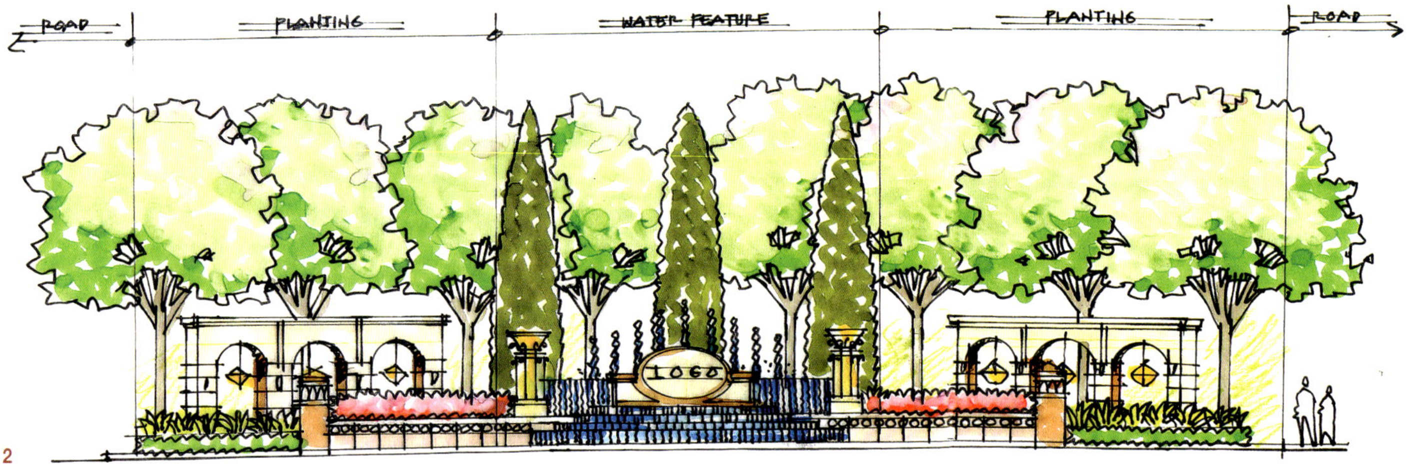

2

BASIC SITUATION

The project is located in the Jiuhua industrial zone of Xiangtan city. The 30.9 hectares land is 370-549 meter long from east to the west and 680-730 meters long from north to the south. There are roads at its 4 ends. The land is 26 km away from Changsha, capital city of Hunan province, or 18 minutes car drive. The accessible land has a perfect base in terms of the ecology, air, mountain shape, plants and the natural environment.

AN ANALYSIS ON THE LAND

The basic characters of the land include 10-15 meter altitude, two 8,000m^2 waters and low mountains and valleys. The design manages to maintain the natural elements, making the project a top-end villa garden. The 7 villa complexes, two castle chambers, two impressionism theme waters as well as the steeple, circular arch, columniation, fountain, square, sculpture and courtyard help impose the European royal verve into the project.

A survey finds the following positive and negative charters of the project

3

4

5 6

7

Positive elements

The existing buildings are well arranged, imposing little impact to the image of the project.

The two lakes with different altitudes and shapes are good for providing differed and systematic sceneries.

Negative elements

The two lakes have an altitude difference ranging from 1.5 to 7.7 meters. This is not a good news for the privacy of the villas and creation of waterside sceneries.

In front of the villa chamber is a 40 meters long and 17 percent ramp. It is a headache for the scenery arrangement and firefighting. Partial villas have little grassland for arranging the matching houses.

DESIGN POSITIONING

The scenery needs to stick to the style in European countries and the US to echo the building style. The method is to cut the project into different areas, making the sceneries varied and of the same level.

1. 湖畔花园中心区剖面图
2. 逸墅广场立面图
3. 芭蕾花园中心区
4. 迷迭香花园
5. 天使堡景区
6. 拾梦桥及小岛水景区
7. 莫奈大街

1

2

工程档案

工程名称
湘潭逸墅庄园景观设计

工程地点
湖南省湘潭市

设计单位
湖南省建筑设计院景观设计研究所

设计师
王小保、彭丹、唐贤巩、简天佐、张剑飞、许海南、范米娜、戴欣、沈蓓

项目景观总占地
47公顷

设计完成时间
2007年9月

竣工时间
在建

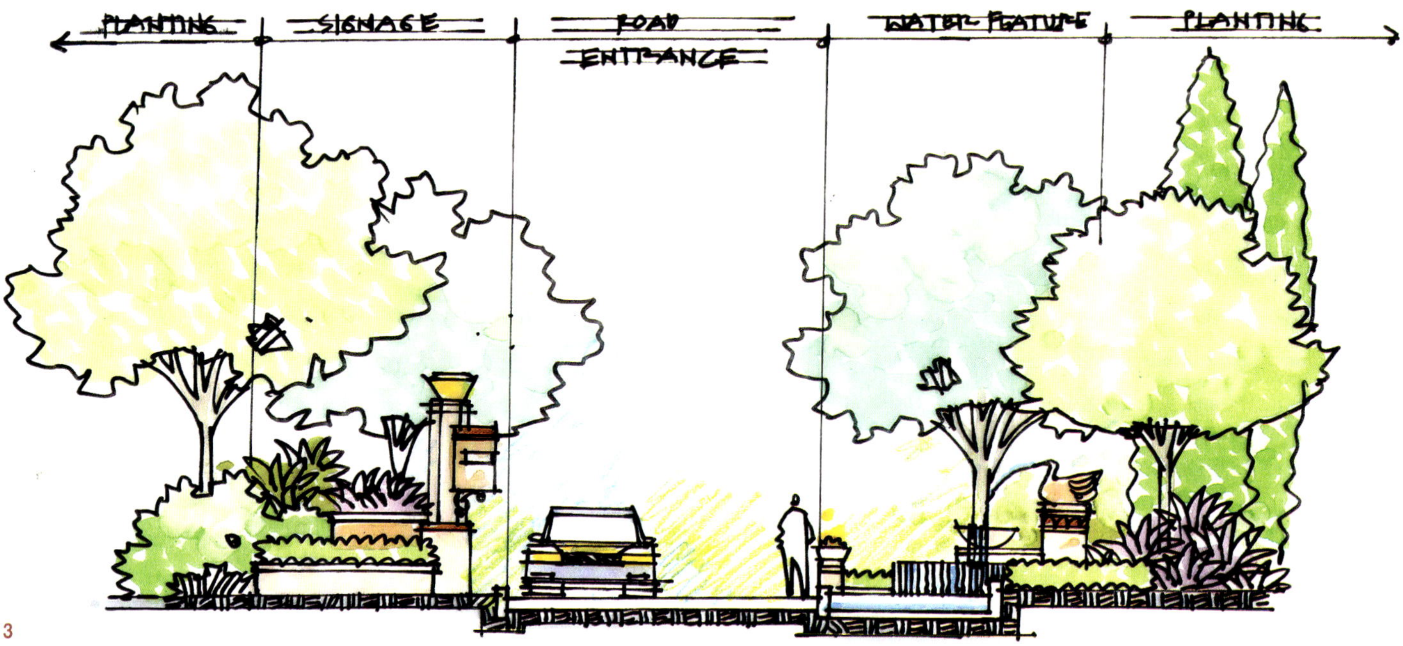

3

1. 天鹅堡跌水墙
2. 逸墅会所中心庭院夜景
3. 波波里花苑北入口
4. 主入口逸墅广场平面图
5. 天使堡

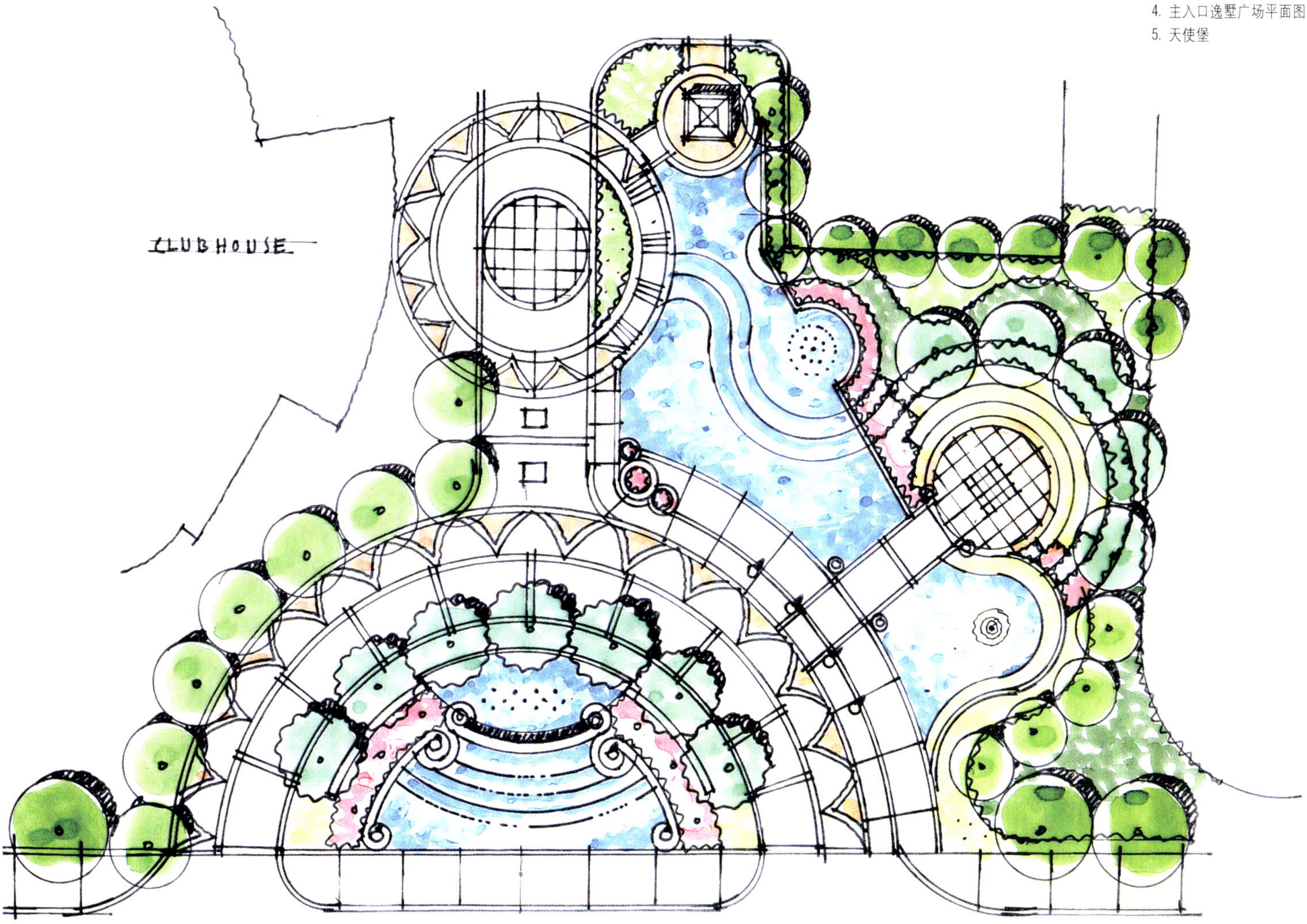

4

5

1

Landscape design of Bandaohuafu, Ningbo city

宁波半岛华府景观设计

设计理念与内涵

宁波半岛华府座落于宁波市近郊，基地北靠河道与城市公园一衣带水，西接繁荣路。本着人与自然和谐、共生的设计原则，半岛华府的景观设计将体现"Natural Healthy Sustainable Communities"的概念，即人类的健康和自然生态系环境之间密切关系构成人类交流共存的生活方式。本项目设计体现了整体性与多元文化的有机结合；生态性与精神家园和谐共筑；"人——建筑——环境"三者统一，在场所独特的环境优势下，营造出碧水怀抱人间的佳境。

本设计要在概念中探索时间的维度，来体验"完美的一天"：清晨——在清新的空气中休闲健身；上午——在明媚的阳光中进行家庭活动；下午——在宜人的树荫里开展休闲运动；傍晚——在夕阳下放松心情；夜晚——在幽蓝的灯火中寻找自己温暖的家。通过不同时间段生活情景的演绎，使生活在此景中的人们产生"完美的一天"的心理感受。

景观设计

景观空间

总体布局——根据场地的实际情况，在景观设计中把整个场地有机的划分为四个主要景观空间。在现代风格的新古典主义统一下，通过四个景观空间中的景观元素如道路、树阵、水体等有机的穿插和整合，使场地呈现既各具特点又相互渗透的景观布局。整个场地分为四带四块八主题：四带指滨河景观带；四块指四个相互独立又风格统一的居住景

2

3

1. 滨水花园景观
2. 河道两岸景观
3. 亲水码头景观

观空间；八主题指庭院景观、滨水花园、都市森林、宅间客厅、景观大道、人性步道、休闲绿地、多元化主题庭院的八大主题景观空间。

道路景观

"Street side garden"的设计：在主干线道路上用绿色和花卉在不同的地方营造出不同的特征，方便行车，同时也营造了没有树木和花草的组团绿地风景，成为组团居民聚集的场所。道路的边上以生态学的植栽和水景等设计为基础，多数道路都被设计成绿道和园路供行人使用。人行道通过巨大的曲线来提高"绿视率"，道路的宽度尽可能设计的狭窄，道路两侧种植多数为落叶、广叶树种以形成树荫道路。同时在树木的区域规划时不同区域营造不同特色，使人们可以根据树木的不同特色来区分组团。

1. 宅间景观手绘图
2. 半岛华府总平面
3. 木栈道景观
4.5.6. 宅间景观

工程档案

工程名称
宁波半岛华府景观设计

工程地点
浙江省宁波市

设计单位
中国美术学院风景建筑设计研究院
现代设计研究所

设计师
赵阳、钟碧涛、章建明、毛迎荣

项目景观总占地
5.2公顷

单位平米造价
450元

设计完成时间
2006年12月

竣工时间
2007年11月

3

4

5

6

DESIGN PHILOSOPHY AND CONNOTATION

The project is located at the outskirt of Ningbo city, neighboring the riverbed, city garden and Fanrong road in its northern and western sides. The design is expected to reflect the concept of "natural, healthy, and sustainable community". The project highlights a dynamic combination of the entire and multi-end culture as well as an integration of people-building-the environment.

The design tries to offer the residence a "perfect day". Early morning-to do physical exercise in the fresh air; forenoon-enjoy family life in the sun shine; afternoon- to engage in leisure activities under the tree shadow; dusk-to get relaxed under the setting sun; night-to find home in the lightings.

1. 局部剖面图
2. 生态岛景观
3. 景观桥正面
4. 半岛华府主入口
5. 河道两岸景观

1

下 跨页

1. 亲水平台景观
2. 景观桥两岸
3. 河道两岸景观
4. 观景台
5. 林荫道

2

3

4

SCENERY DESIGN

Scenery space

Overall layout-the project is cut into 4 scenery areas. The new classic style dominates the elements like road, tree group and waters, making them peculiar and penetrative to each others.

The entire project is cut into isolated parts and each of them has a theme, such as courtyard, riverside garden, city forest, scenery avenue, walking path, grass land, etc.

Road scenery

The street side garden concept is introduced in the design. The green color and flowers are employed alongside the main roads and places without plants are designed for the residence collection. Most roads are designed for both greening path and garden road. The pedestrians are made curve as much as possible for the walking people to notice more trees. Trees alongside the roads help create greater shadow. The residence can identity the different complexes from the color of the tress inside the community.

5

1

2

3

4

5

The landscape of Xiamen paradiso

厦门万科金域蓝湾景观设计

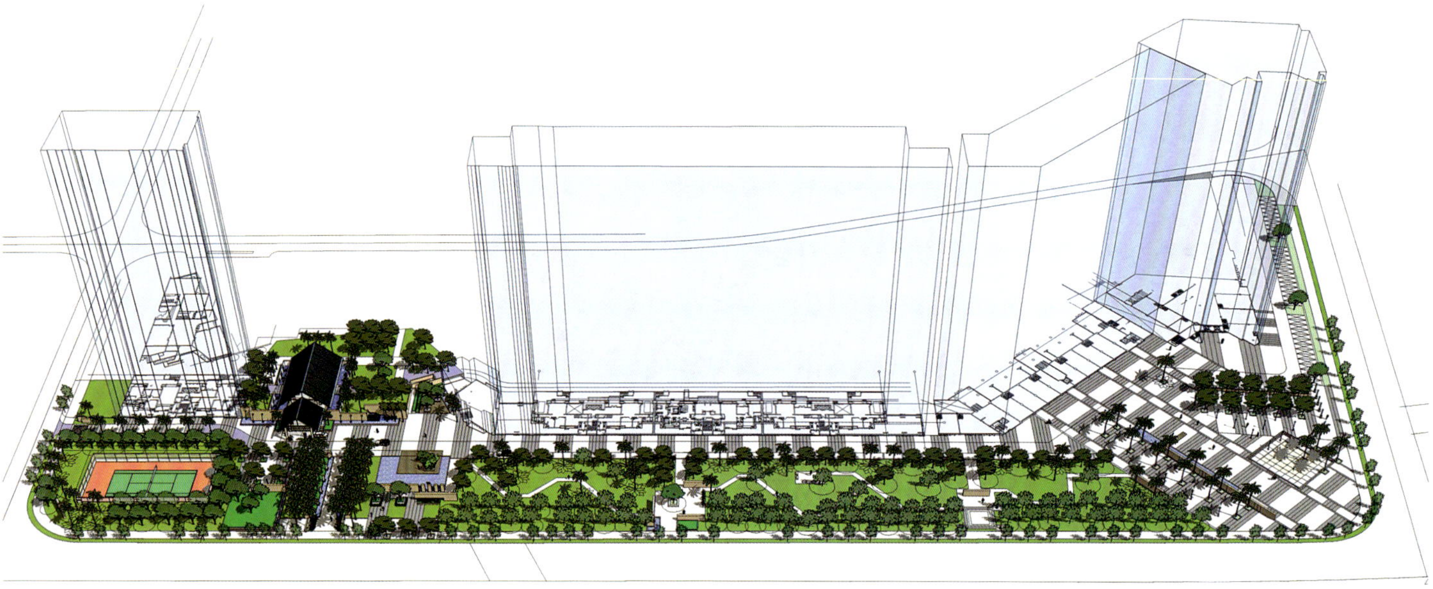

厦门金域蓝湾项目北邻吕岭路，西面接洪文路;南侧与西侧的规划道路将上述两条市政道路贯通，将项目A、B地块分隔开来。这样的平面布局方式一方面不利于项目A、B地块空间上的贯通与连接，另一方面又有利于加强项目区域外的联系，对于项目A、B地块的外围商业起到一个有力的促进作用。

小区主入口区域由一个相对紧凑的轴线空间作为引导，将观者引入小区的入口洗礼区域。在设计轴线上考虑到景观移步异景的效果和人对事物猎奇的心理，将市政入口轴线与小区入口轴线做了适度的错位处理。人们通过一个紧凑狭长的空间看到的对景是一组有万科项目LOGO的喷水景墙，走出这一区域眼前豁然开朗，小区入口洗礼区就出现在眼前，一张一弛、一开一合中体现了东南亚风情园林的趣味所在，利用景墙，营造出"收、放、开、合"的绿化空间。

考虑到市政设计对绿化覆盖率60%～70%的设计要求，我们将这个区域定义为集中的绿化区域，在保留原有市政人行道及两排凤凰木的前提下，增加绿化面积及大型乔木的种植空间。只有两处设有尺度适宜的人行入口，与绿化带中的人行步道连接，形成漫步体系。考虑到绿化带与项目商业街相连，在设计中尽量减少中层次的灌木渲染，转而强调草坪与大型乔木形成的开阔型视线空间，一方面保证市政绿化带在视觉层次上的丰富性，另一方面也使得商业界面与外界在空间上是通透的、联系的。 商业广场区域是以硬质铺装为主的开敞商业空间，在设计中以有涌泉肌理的带状水体作为引导，以景墙与热带植物作为映衬，将人流引导向项目的商业空间，以矩形旱喷下沉广场和枝干有较强肌理的棕榈科植物矩阵穿插结合的景观，丰富广场景观层次，再将遮阳伞与休闲座椅设置其间，使得广场景观在设计内容上更为丰满。

厦门金域蓝湾以"关注城市文化，建筑人文生活"为主旨，提倡一种全景观阳光住宅，强调对居住空间的利用，将人的活动空间塑造成一个全阳光、全景观、全通风的环境，力求打造成一个集自然环境优势、生活配套完备、城市文化汇聚、新兴科技集纳的全新城市生态景观高尚居住区。

1. 主入口
2. 游泳池休闲区

2

吕 岭 路
箱涵边线
污水管
People Main Entance
Buffer Trees
Feature Wall
Palm Trees
Dry Fountain Plaza
Fountain
Feature Paving
Commercial Plaza
Umbrella
Commercial Street
Parking Space
Commercial
Foot Path
Planting
Water Feature
Wooden Deck
Lawn Steps
Tenis Court
Sculpture
Main Road Inside Group
Sale Management Center
Christening Entrance
Adoult Pool
Childrens Pool
Foot Wash
Club House
Kidergarten
Kids Playground
Pavilion
Tree Box
Underground Parking Space Entry
People Entry
Red Line
规划道路
洪文路
EVA

The Xiamen paradiso project is adjacent to Lvling road and it connects Hongwen road in the south. The planned roads in the south and west sides will connect the above-mentioned two roads while cutting the project A and B. The consequence is in one hand not good for connecting the project A and B in terms of the space, and in the other hand, it will help strengthen the outer connections of the two projects, in terms of the commercial advantages.

The main entrance area is led by an axial space which is comparatively compact and visitors are lured to the baptism area in the entrance. In designing the axial lines, full considerations were paid to the change of the sceneries of a moving person and people's mind of witnessing novelty. The two entrance axial lines were intentionally misplaced. After passing a compact and narrow space, what faces people is a water spray scene wall with a logo of Wanke. People will be suddenly enlightened walking out of the area. The contrast embodies the interesting part of southeast Asian style and the wall creates a green space of collection, relax, open and shut.

Because the urban planning authorities demand the greening proportion of the area be no less than 60~70 percent, the designer, while keeping the pedestrian roads and two rows of flame tree, increases the amount of greening areas and large-sized arbor. There are two walking entrance which connects the pedestrian walk inside the greening areas for walk use. Because the greening areas is connected to the commercial street, the designers have limited the use of mid-level shrubs, and highlighted the wide vision space forged by grass land and large-sized arbor. The design has managed to maintain the vision continuity of urban greening areas and made the commercial and outer areas throughout and connected in terms of the space. The commercial plaza is an open place supported by hard brick floor. Boiling spring shaped water is introduced in the design and is set off by scene wall and tropical zone plantings. Visitors are lured to the commercial area. Rectangle sink plazas and palm plantings were used alongside the line. Sun shades and chairs help make the scene more opulent.

The project, under theme of paying attention to city culture and building humane life, advocates full scenery sunny house, highlights the use of living space. It tries to build a high-end community areas integrating the natural advantages, living facilities, city cultures and high techs.

1. 儿童泳池景观柱
2. 主入口特色水景
3. 总平面
4. 主入口喷水雕塑
5. 主入口喷水景观

工程档案

工程名称
厦门金域蓝湾景观设计

工程地点
福建省厦门市

设计单位
SED新西林景观国际

设计师
黄剑锋、李昆

项目景观总占地
5.6公顷

设计完成时间
2007年8月

竣工时间
在建

1. 主入口构筑廊架
2. 泳池区喷水雕塑
3. 主入口特色水景
4. 主入口
5. 主入口商业街
6. 主入口景观

万科·金域蓝湾
the paradiso
5

万科·金域蓝湾
the paradiso
6

Landscape design of Foshan Vanke Rancho Santa Fe Residence

佛山万科兰乔圣菲高尚居住区景观设计

在佛山万科兰乔圣菲居住区景观设计中，首先设计师以研究居民的日常行为和生活体验为第一要务。从入口人车分流的管理、道路尺度的控制，到会所前广场的开放空间、泳池的各项使用、再到活动场地的安排，高层架空层的使用，始终坚持着功能优先，以促进住户交往为宗旨。另一方面，设计师也在坚持着景观美学的追求，从概念到细节，从材料到色彩，努力体现着西班牙式的纯粹风格。环境景观的品质提升了使用者的美感和舒适度，也就在潜移默化间改变和影响着居民的生活。

小镇会所是西班牙式建筑，建筑外立面色彩明快，采用质朴温暖的色彩，充满了阳光的地中海味道。以社区会所为背景的中心水景区域，是景观设计的重点。对称的种植形式（大王椰子、凤凰木）将人们的视线引导到中心水景区域。这个区域里热情奔放的地面铺装色彩、活泼动感的喷水景观与自然而浓密的油棕结合在一起，相映成趣。共同构筑了西班牙式的主入口景观。

观者由主入口进入行至水景区便可影影绰绰的看到远处的风情会所，经过中心水景进入到会所区域，

1. 露天游池
2.5.6. 主入口水景区
3. 会所区域西班牙欧式建筑
4. 道路宅间

1

2

3

4

6

In the design of Rancho Santa Fe project, designers managed to make the function facilities more convenient for the residents to communicate to each others.

Meanwhile, the designers pursue a perfect Spanish style in terms of the concept, materials and colors.

The town chamber is of Spanish style, reminding people the Mediterranean sea. The water scenery area is focus of the design. Symmetrical plants lure people's vision to this area. Other distinctive features include floor bricks, fountains and oil palm.

Townhouse: each unit has two yards, for guests and family use respectively. The thick walls and round corners make the residents feel comfortable.

The entrance floor materials are full of Spanish style too. The scenery walls, slates and scree are typical Spanish factors.

The residential area also introduces a lot of typical Spanish plants.

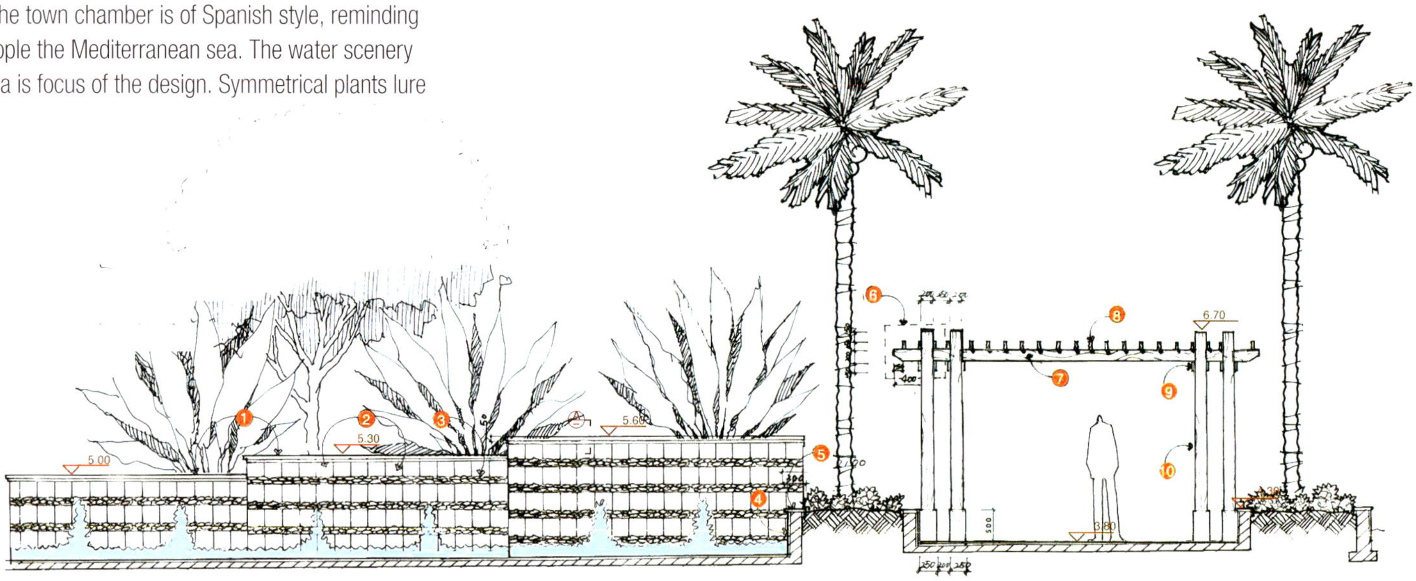

1

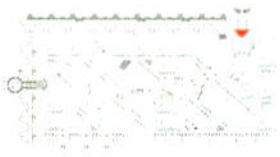

2

3

4

工程档案

工程名称
佛山万科兰乔圣菲高尚居住区景观设计

项目地点
广东省佛山市

设计单位
SED新西林景观国际

设计师
黄剑锋、李昆

项目景观总占地
13公顷

设计完成时间
2006年8月

竣工时间
2007年5月(一、二期建成)
三期在建

1.5. 庭院一角
2. 主入口水景效果图
3. 休闲区
4. 中心水景效果图
6. 联排别墅区
7. 入口
8. 庭院入口
9. 社区小品

5

6

8

9

7

Landscape design of Zhongxin Chaohuayaju (second phase), Shantou city

汕头中信潮华雅居二期景观设计

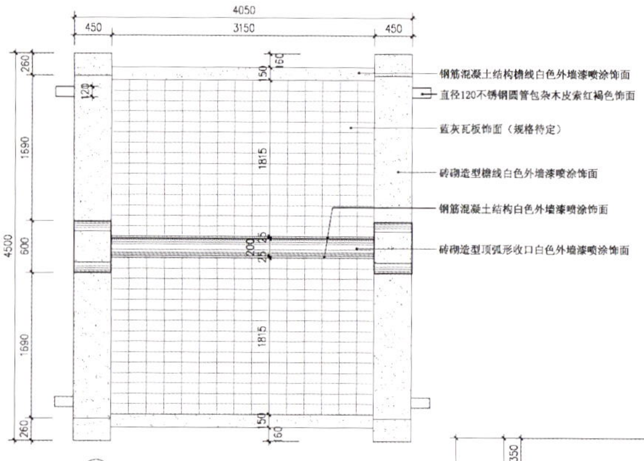

1 湖心赏鱼台亭顶面平面图 1:30

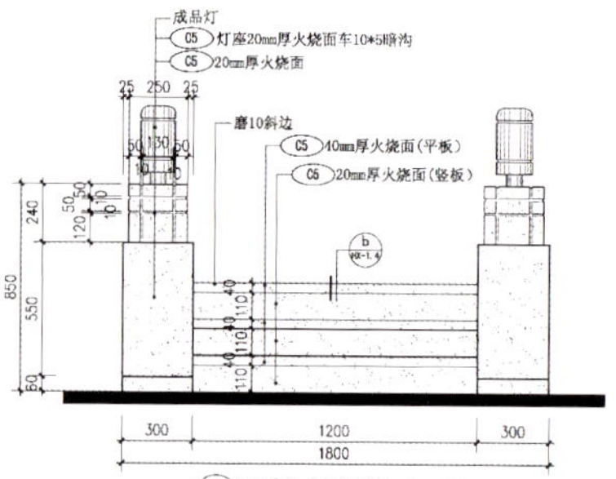

A 石桥立面图 1:15

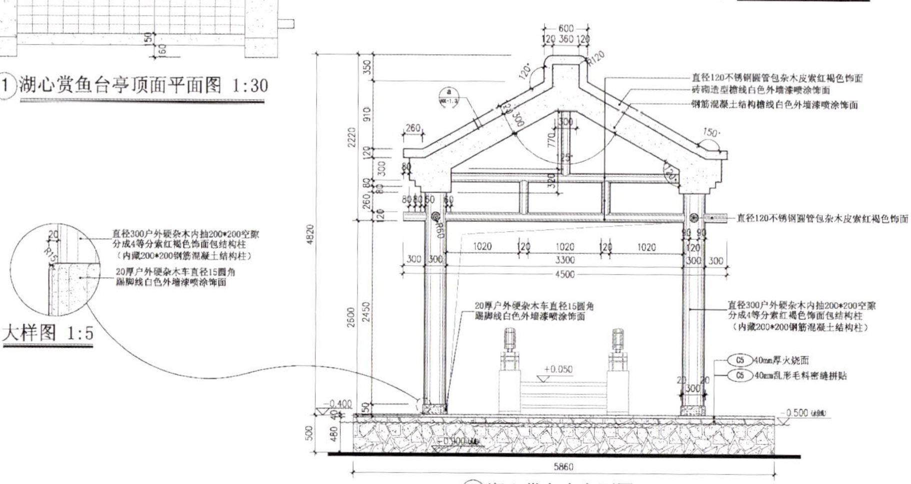

大样图 1:5

B 湖心赏鱼台立面图 1:30

注：钢筋混凝土结构根据高度及位置预埋铁板可供给不锈钢圆管焊接（做法由施工人员自定）。

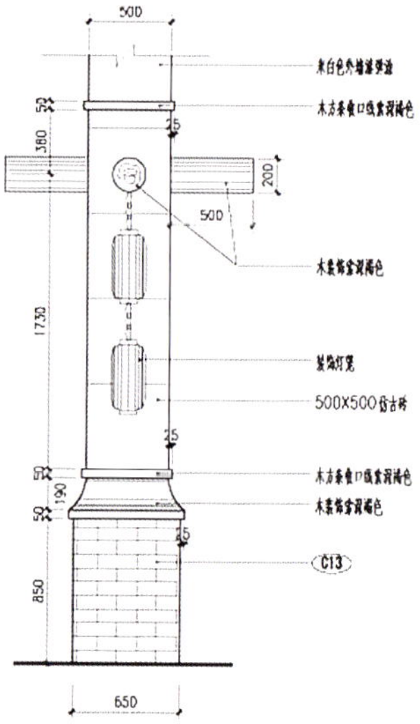

装饰柱立面图 1:20

“潮华雅居”的小区景观设计着力打造一个富含潮汕传统民间生活气息的“情景空间”，将演绎着与潮居生活息息相关的小桥流水、莲花、芭蕉、榕树、石臼、蓑衣、风车、木雕、石刻等作为小品点缀其中，带给我们心灵的温馨，为小区住户营造和睦相处、其乐融融的生活空间。小区显得温情而轻松，每个梯间照壁上的薄雕，采用民俗题材，以燕筑巢、竹、牡丹等喜庆吉祥图案，用福建青石精心打制而成，精美而生动，以独特的方式展现潮人的精神追求和道德礼仪。“花开富贵”“龙凤呈祥”“桃园结义”“郭子仪拜寿”，一屏一故事……小区内像是不经意地散落着其他天然石块，有的形如骆驼，有的状若鸳鸯……皆是惟妙惟肖，而且点缀得恰到好处。

潮华雅居因其建筑颜色采用潮汕民居传统的青灰色，因此在进行园林设计时考虑了大量各种颜色的花木。春季，桃花怒放，亭亭玉立的美人树上绽放着满树的花朵，满园的玫瑰也竞相绽放。夏季，大叶紫薇，小叶紫薇，粉叶决明，荷花、龙船花，鸡冠刺桐，栀子花、茉莉花，将夏天

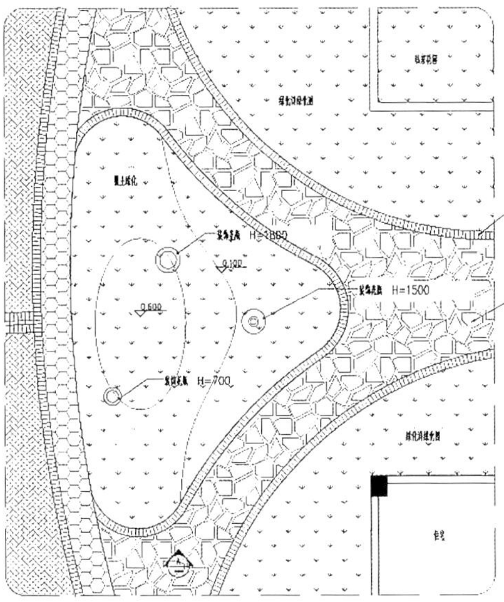

街心绿化区平面图 1:50

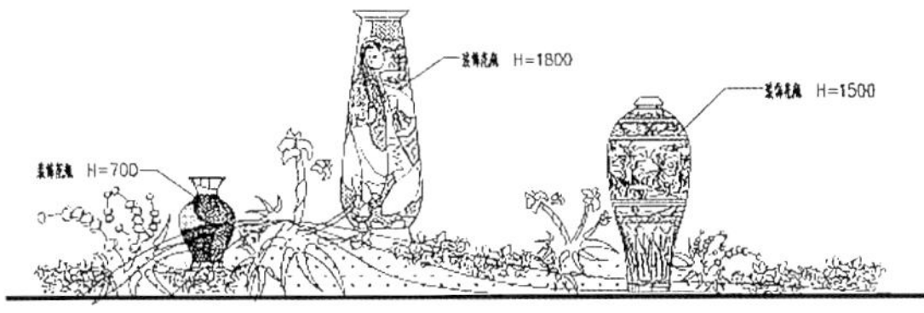

Ⓐ街心绿化区立面图 1:25

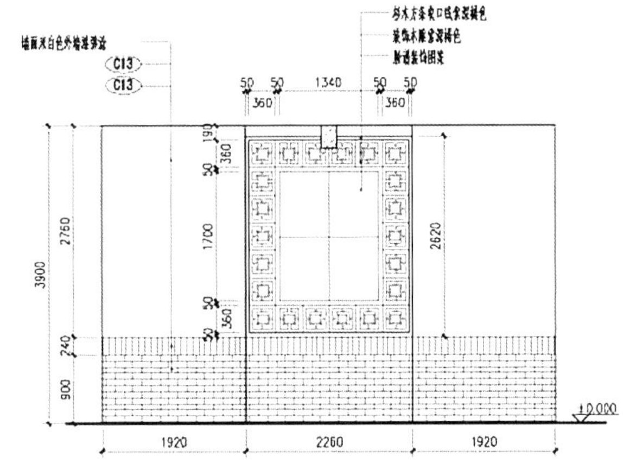

Ⓒ架空层立面图 1:50

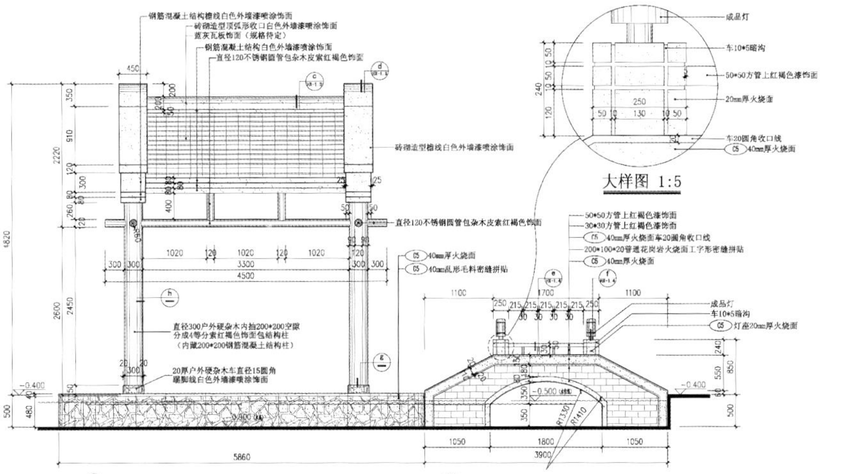

大样图 1:5

Ⓒ湖心赏鱼台立面图 1:30

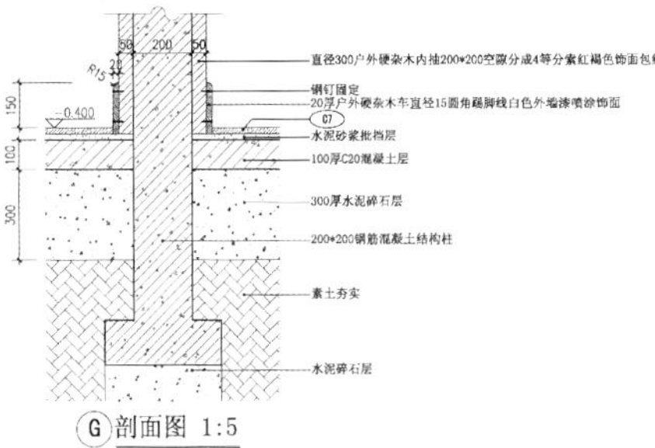

Ⓖ剖面图 1:5

注：柱脚根据现场实际情况由施工人员自定。

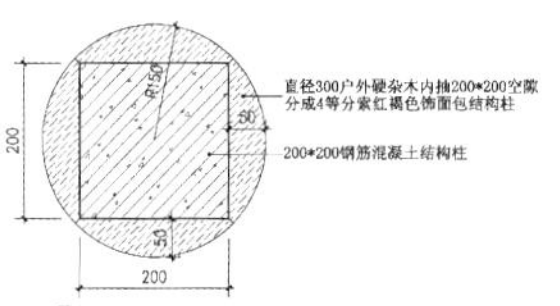

ⓗ剖面图 1:5

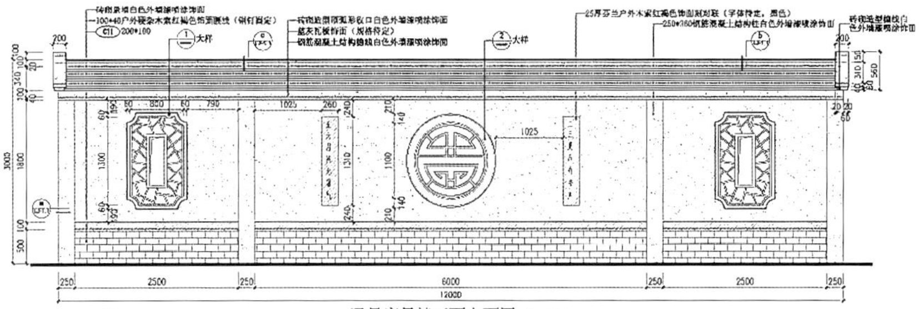

漏景窗景墙正面立面图 1:30

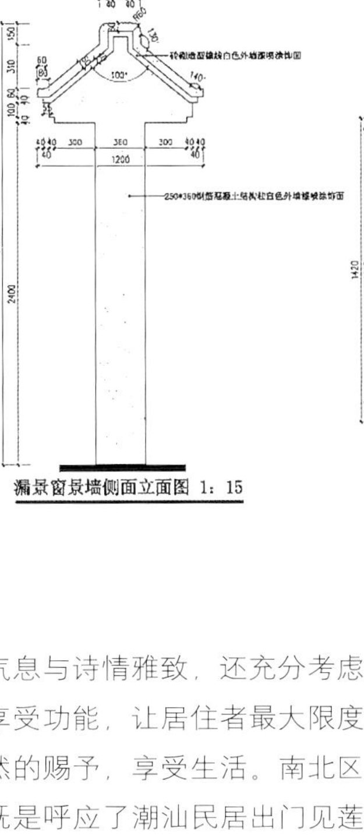

漏景窗景墙侧面立面图 1:15

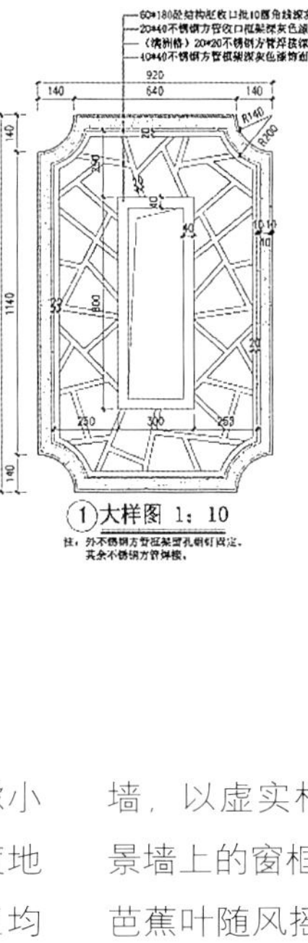

①大样图 1:10

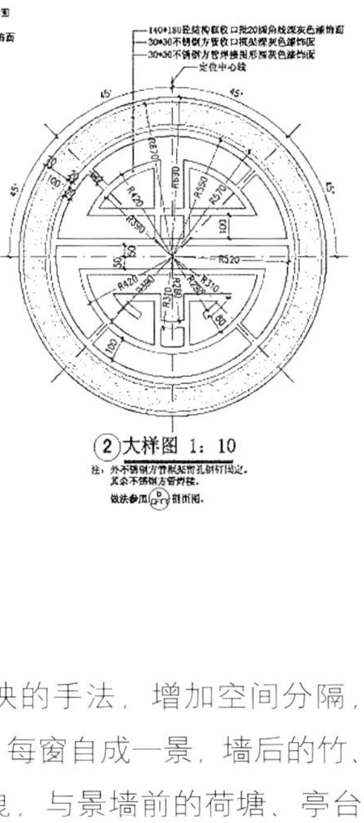

②大样图 1:10

的潮华打扮得五颜六色，清新怡人。秋季，虾衣花、桂花、米兰、木芙蓉也给潮华雅居送来阵阵清香。冬季，满园的角花给寒冷的冬天带来温暖的气息，设计者在园林规划上不仅重视视觉效果，更贴心地考虑到居住者的嗅觉效果，以期达到赏心悦目的效果。

在园林和环境的布置上，不仅强调楼盘的文化气息与诗情雅致，还充分考虑小区的生活享受功能，让居住者最大限度地利用大自然的赐予，享受生活。南北区均有水体，既是呼应了潮汕民居出门见莲池的地方特色，又为居住者提供赏荷观鱼的休憩空间，更有调节小环境气温的功用。在北区，水面为主庭院，亭子建在水中或岸边，成为水榭。另外还特别设了一道景墙，以虚实相映的手法，增加空间分隔，景墙上的窗框，每窗自成一景，墙后的竹、芭蕉叶随风摇曳，与景墙前的荷塘、亭台共同构筑一个愉悦轻松的场景，景中有景，幽深广阔。南区则是水随园转，荷风四面，杨柳轻垂，极具唐诗宋词之深远意境。

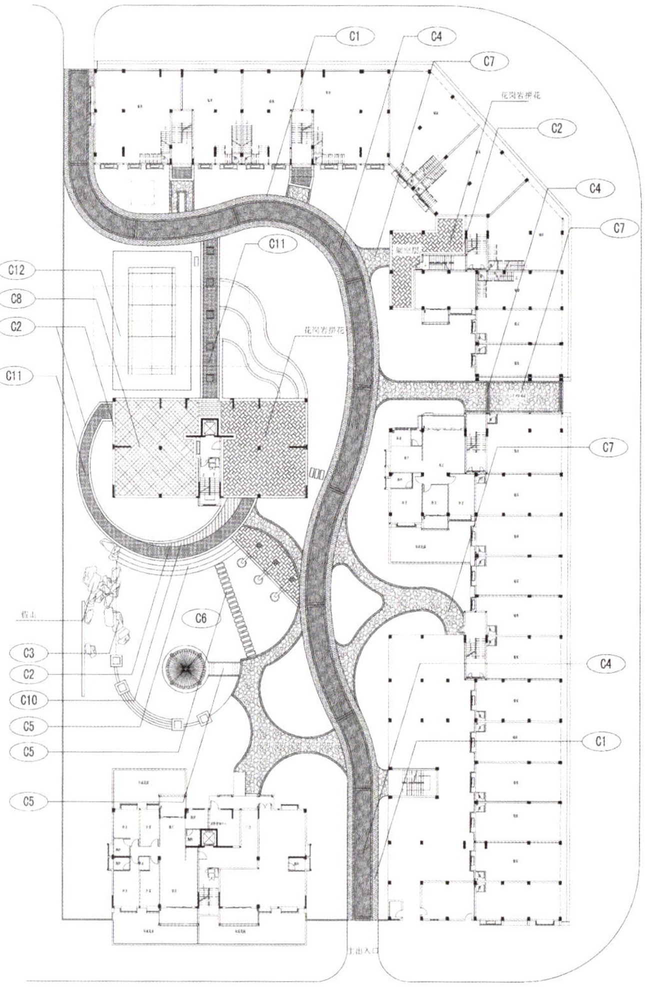

工程档案

工程名称
汕头中信潮华雅居二期景观设计

工程地点
广东省汕头市

设计单位
汕头市天遨环境装饰设计有限公司

设计师
陈楠、陈灿明、谢贵彬、王海元、
林书概、杨金敏

项目景观总占地
3.56公顷

单位平米造价
280元

设计完成时间
2006年1月

竣工时间
2007年1月

1. 湖心亭
2. 小桥流水
3. 情趣小道
4. 阳光下的小区道路
5. 虚实相映中的景墙
6. 荷塘
7. 翠绿中的文化小品

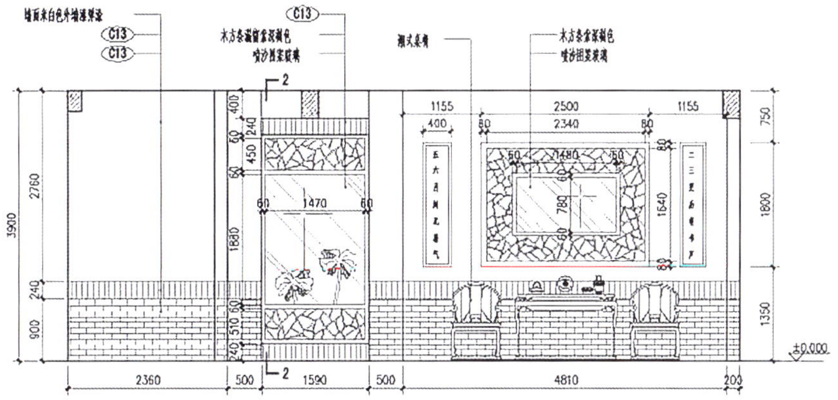

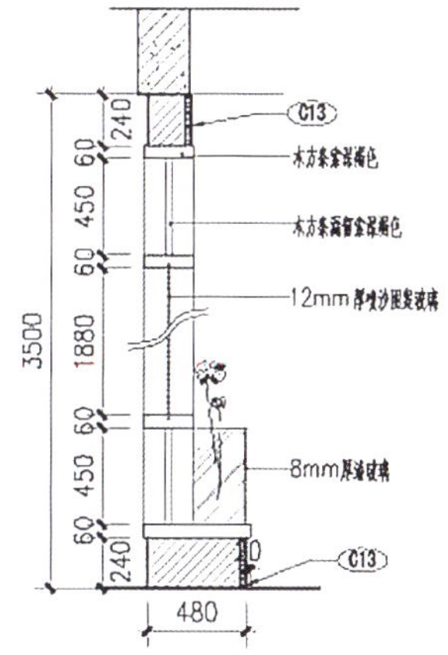

2－2剖面图　1:20

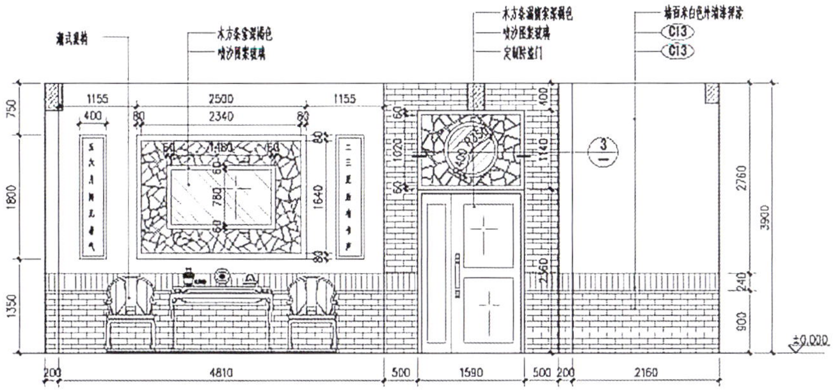

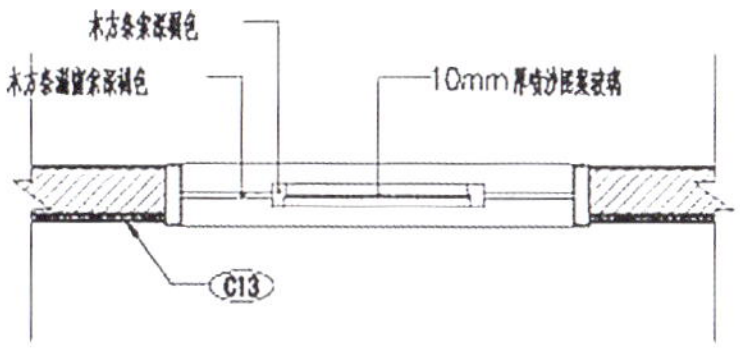

3－3剖面图　1:20

The project aims at creating a kind of scenery space that is full of the folk living breath of Chaozhou-Shantou, making use of the things such as bridge, lotus, banana, banian, stone mortar, coir raincoat, windmill, woodcarving and carved stone in the residential community. Sculptures on each screen wall come from folk-custom stories, illustrated by the lucky patterns of incubating swallow, bamboo and peony. The patterns are made of the Qingshi coming from Fujian province. Each wall carries an individual historical story. Also dropped in the community are stones look like animals such as camel and mandarin duck.

Colorful plants are chosen to be contrast to the grey color the community adopts. The plants make the community gorgeous in the four seasons. The designers pay attention to the olfactory sensation of the residents as well.

The designers try their best to make the residents enjoy the nature. There are bodies of water in the south and north ends of the community. In the north end, the water body is the main courtyard, with pavilions being built inside or alongside the area. The water body in the south yard is changing with the garden.

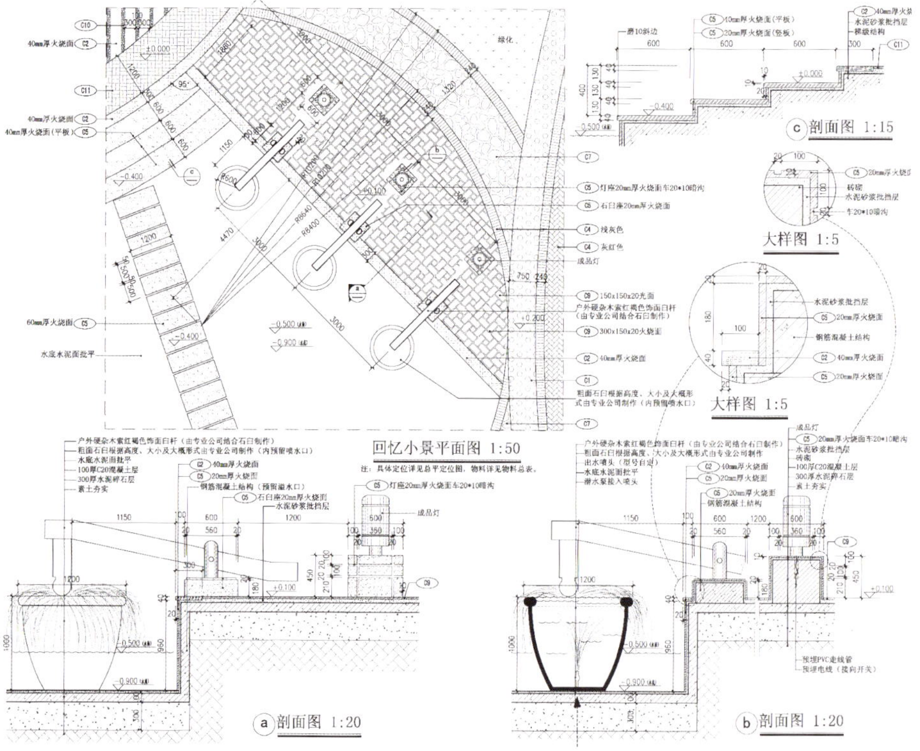

1. 照壁薄雕与建筑融为一体
2. 景观小桥
3. 曲径通幽
4. 翠绿中的文化小品
5. 曲溪
6. 假山流水

Landscape design of Fulitaoyuan residential community, Beijing

北京富力桃园景观设计

设计理念

本项目景观设计以"现代、人文、艺术、原生"为四大主题，着力营造具有浓郁现代气息的生态景观环境，满足都市人对原生态景观的追求与渴望，全面提刀社区的品位与档次。

总体景观布局

四大分区

项目总体分为四大区，分别为A、B、C、D区。其中A、B区为一期建设区(亦为景观示范区)，C区为二期建设区，D区为三期建设区。外围绿化带是小区与外环境的隔离区域。

四大主题

现代——设计中运用多种元素体现这一设计主题：新型材料的运用；流畅、简洁的设计手法；现代化景观建筑和小品的运用等。

人文——充分挖掘景观的文化内涵，使人们在视觉享受的同时，深层次的精神文化需求也能得到满足。

艺术——抽象的雕塑、不同形式的景观建筑构架、各种形态的景观小品……它们是跳跃的音符、流动的韵律。

原生——扩大并强化场地原有湿地，保护重要的生态环境，创造出现代钢筋水泥城市中少见的原生态湿地景观，形成整个小区的绿色核心区域，赋予小区回归自然、可持续的循环生态系统。

一大主要特色——开阔、丰富的水体景观

整体景观有一万多平方米的宽广水面，考虑到水体难以维护，设计中心水面采用了具有自净能力的湿地水体，既减少后期维护成本又形成奇特的湿地景观。设计师在设计中运用多种对比：一处宽广中心水面与多处精巧的组团水体；大型的湿地水面景观、自然卵石铺底的水体与蜿蜒曲折的溪水水道；自然、原生的湿地水体与规则、现代的涌泉、叠水等人工成分较大的水体……这些对比无不体现水体景观的丰富性、多样性、艺术性。设计考虑了生态、人性的各个方面，极力营造宜人的水体环境，满足人亲水的天性，最大程度地让人们回归自然。

两大功能分区

小区总体上分为区内景观区和区外休闲区。区内景观区包含了示范区、艺术欣赏区、运动区、读书区、休息区、儿童游乐区、原生态水景区、商业区、会所景观区。区内景观区像一个个珍珠撒在景观平面，构成丰富的景观功能独立点，同时各区之间相互衔接、渗透，处处体现现代、人文、艺术、原生这四大主题。

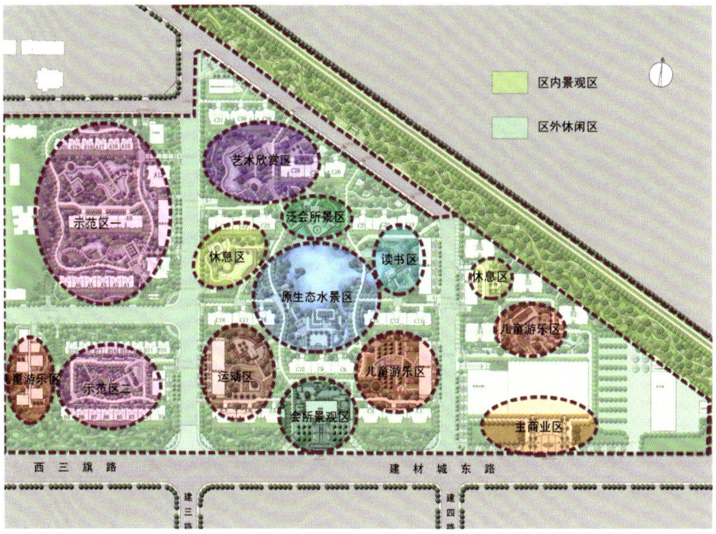

DESIGN PHILOSOPHY

The design highlights the themes of "modernity, human, arts and nature-made". The ecology-based sceneries satisfy people and help upgrade the entire project.

ENTIRE SCENERY LAYOUT

4 sections

The entire community plans to be finished in 4 phases and a greening belt separates it from the outside world.

4 themes

Modernity-multi elements have been used to show the themes, including new materials, fluent and compact design methods, modern landscape buildings and ornaments.

Human-make full play of the cultural connotation, providing people with visional enjoyment and in-depth emotional touches.

Arts-there are sculptures, buildings and ornaments.

Nature made-to extend the area of wetland, a scarcity in an area surrounded by modern buildings. It is the focus of the green community, providing a nature-oriented and sustainable eco system.

A wide wetland

The wetland water can make it clean by itself; multi contrast approaches have been used and the contrasts take full considerations of the nature and human.

Two main functional divisions

The inner landscape area offers functions for entertainment of the residents and each area is connected to the others.

工程档案

工程名称
北京富力桃园景观设计

工程地点
北京市海淀区

设计单位
广州市太合景观设计有限公司

设计师
赵伟强

项目景观总占地
13.3公顷

设计完成时间
2006年9月

竣工时间
在建

中国建筑设计研究院环艺院景观所

LANDSCAPE DESIGN DEPT. of ENVIRONMENTAL ART DESIGN & RESEARCH INSTITUTE, CHINA ARCHITECTURE DESIGN & RESEARCH GROUP

■ 企业介绍 INTRODUCTION

中国建筑设计研究院环艺院景观所是以园林景观规划设计为主的具有甲级设计资质的综合设计所，通过ISO9001国际质量体系认证。业务范围包括：风景区、旅游度假区、主题公园的规划设计；城市设计；城市公园、广场、商业街的景观规划与设计；建筑周边环境景观设计；居住小区景观规划设计等。

景观所目前有专业技术人员30余人，其中中级以上职称占60%，并拥有全国注册城市规划师， 国家一级注册建筑师等专业技术人才。在与业主、规划师、建筑师的配合方面有着丰富的经验和较大的优势。

中国建设文化艺术协会环境艺术委员会常务理事单位。主办《中国景观设计年刊》。协办《景观设计》杂志， 参编国家标准图集《环境景观室外工程细部构造》、《环境景观绿化种植》、《建筑场地园林景观设计深度图样》、《种植屋面建筑构造》等。

景观所本着“质量第一、信誉至上”的宗旨，并充分利用中国建筑设计研究院的科研设计、技术人力、资产结构等方面的优势，多年来一直与业主、规划师、建筑师等保持着密切的合作关系，完成了多项高质量的景观规划设计任务。在稳固发展的同时，兼顾社会义务，积极探索、努力创新，为使我们的生活更美好而不懈追求。

■ 主要业绩 MAIN PROJECT

主题公园景观规划：
奥林匹克森林公园及中心区景观规划设计
布达拉宫周边环境整治及宗角禄康公园改造规划设计
北京回龙观居住区文化公园景观规划设计
南京佛手湖 中国国际建筑艺术实践展景观规划设计
无锡鸿山遗址公园博物馆景区景观设计

城市设计、城市广场：
首都国际机场三号楼楼前景观与城市规划
奥运公园观景塔景观设计
北京西单商业区城市设计
莫斯科中国贸易中心景观设计
凉山民族文化艺术中心及火把广场景观设计
重庆解放碑广场景观设计
西安南门广场景观设计
衡水市城市南入口及西入口规划设计
中国农业博物馆景观规划
苏州火车站站前广场景观设计
山东东营清风湖商业街景观设计

公建及其周边：
奥运会国家主体育场景观设计
中国驻南非大使馆景观设计
首都博物馆新馆外环境景观设计
拉萨火车站北广场景观设计
北京外研社国际会议中心景观设计
北京德胜尚城景观设计
北京韩美林艺术博物馆景观设计
大连软件园系列景观设计
北京数字出版信息中心景观设计
北京西直门交通枢纽景观设计
山东广播电视中心景观设计
安徽出版局景观设计
北京地坛医院迁建工程景观设计

校园景观设计：
北大留学生公寓景观设计
北京外国语大学校园景观设计
唐山学院校园景观设计
北京体育大学校园景观设计
北京四中校园景观设计

旅游度假、滨水景观：
月亮峡谷旅游渡假区景观设计
北京凉水河滨河绿带景观设计
廊坊市龙河及环城水系景观规划设计

居住小区：
北京万科紫台项目景观设计
北京万科石佛营项目景观设计
北京万科西山庭院景观改造
鞍山万科金色家园景观设计
沈阳万科魅力之城景观设计
北京万科公园五号居住区景观设计
北京万科东第项目景观改造
北京万科假日风景景观设计
山东临朐滨河花园景观设计
山东东营金瀚家园环境景观设计
南京聚福园景观设计
济南汇统花园景观设计
湖南常德康桥蓝湾小区景观设计

中国建筑设计研究院 环境艺术设计研究院 景观所
China Architecture Design & Research Group
Environmentral Art Design & Research Institute
Landscape Architecture Dept.
电话（Tel）：86-10-68302745/68302780
传真(Fax)：86-10-68360982/68361229
E-mail:jgs@cadg.cn
地址：北京市车公庄大街19号二号楼九层
邮编：100044
Adress:NO.19 Che Gong Zhuang St.Beijing
100044.P.R.China

无俗念
烂银霞照通彻
浑似姑射真人
天姿灵秀
意气殊高洁
万蕊参差谁信道
不与群芳同列
浩气清英
仙才卓荦
下土难分别
瑶台归去
洞天方看清绝

中国建筑设计研究院环艺院景观所

辽宁公安司法管理干部学院

河南国家大学科技园

新疆鄯善县

中山岐江公园

秦皇岛汤河公园

沈阳建筑大学校园

浙江永宁公园

建筑

城市

土地

土人设计
TURENSCAPE

北京 . 上海 . 广

北京土人(土人设计
地址：北京市海淀区上地信息路
中关村发展大厦一层(100
北京土人景观与建筑规划设计研
电话 010-62988907 010-6297
传真 010-62988905 010-6296
www.turenscape.
E-mail: turen@turenscape.

十年锤炼，成就世界品牌
天地人神，引领国际潮流
志同道合，梧栖天下才俊
至精至诚，倾心品质服务

北京土人景观与建筑规划设计研究院(简称"土人设计")，由美国哈佛大学设计学博士、北京大学教授俞孔坚领衔创立，拥有300多名职业设计师，由归国留学人员构成设计主体。为综合型、国家甲级资质规划设计单位，北京市高新技术企业，获ISO9001：2000质量体系认证。主要业务范围包括：

- 建筑设计
- 景观规划设计
- 城市规划设计

土人设计以土地的名义，倡导天地、人、神和谐的"土人理念"，坚守品质至上的业道德，高举民族设计大旗，立足本土面向全球。多年来，在国内外完成大量规划设计项目，参与多项国内外重要工程的规划设计包括中关村科技园区、2008北京奥林匹克场馆、2010上海世博园区、北京798艺术区、大运河国家遗产与生态廊道、国土生态安全格局等，先后五次获全美景观设计师协会年度荣誉设计奖，两次获得世界青年建筑师奖，等国际奖项，享誉海内外，引领国际设计潮流。获奖包括：

2007 世界青年建筑师奖 Ar Award Commended (The Architecture Review)
2007 世界滨水设计最高荣誉奖(Excellence on the Waterfront Top Honor Award)
2007 年美国景观设计师协会荣誉设计奖(ASLA Design Honor Award)
2006 年美国景观设计师协会荣誉设计奖(ASLA Design Honor Award)
2005 年美国景观设计师协会荣誉设计奖(ASLA Design Honor Award)
2005 年美国景观设计师协会荣誉规划奖(ASLA Honor Award Planning and Analys
2005 世界青年建筑师奖 Ar+d Award Commended (The Architeclure Review)
2005 中国人居环境范例奖
2004 第十届全国美展金奖
2003 首届全国建筑艺术奖
2002 年美国景观设计师协会荣誉设计奖(ASLA Design Honor Award)
2001 中国首届人居环境范例奖

北京北林地景园林规划设计院有限责任公司的前身为北京林业大学园林规划建筑设计院，有建设部核发的市政公用行业（风景园林）甲级工程设计资质（设计资质证号为建设甲010132-SJ），国家发展和改革委员会核发的工程咨询甲级资质（工咨甲2032101001），是具备规划、设计、咨询等业务能力的专业设计院。

我院成立于八十年代初期。在改革开放之初，我院依托高校雄厚的师资力量，实行产学研相结合，“以研促产”、“以产兴学”。在设计院所深化体制改革之时，我院又率先进入市场，“以质量求生存”，服务社会。不同于其他政府所属的设计院的特殊历程，即造就了我院的基本特色（即：最好地继承和发扬我国古典园林的传统精髓，最广泛地吸取国内外先进的思想理论和设计方法，最体贴地提出符合社会需要的规划设计方案），也练就一支素质优良，专业配套，结构合理，年富力强，勇于开拓的设计队伍。

经过二十年的经营，我院的业务范围不断扩大，编及近三十个省、直辖市。一批涉及规划、设计、施工的获奖、中标项目，如：获建设部2003年度优秀规划设计三等奖的“郑州市老成区居民游绿地体系规划”、“浙江大学新校区环境景观设计”（中标方案）、“郑州市中洲大道景观设计”（国际竞标二等奖）、“首都国际机场3号航站楼环境设计”（国际竞标一等奖）等，确立了我院在同行业设计院所中的地位与影响。

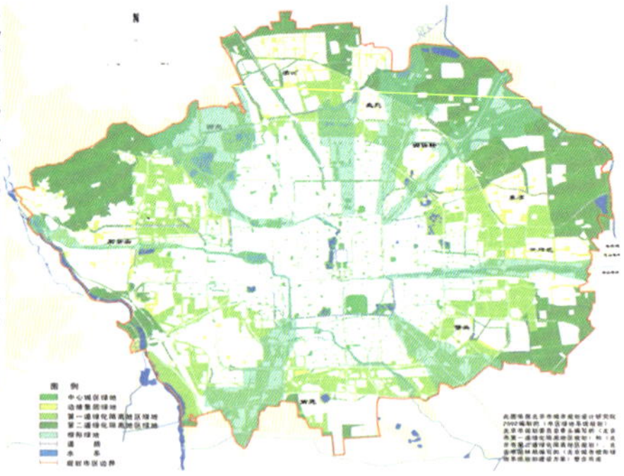
北京战略规划总平面图

海天公园总平面图

北京北林地景园林规划设计院有限责任公司

BeiJing Beilin Landscape Architecture Institute CO.,LTD

地址：中关村东路18号财智国际大厦B-22

电话：（010）82601322 82601588-86

E-mail: bldj@263.net 邮编：100083

网址：www.bldj.com.cn

重庆工学院总平面图

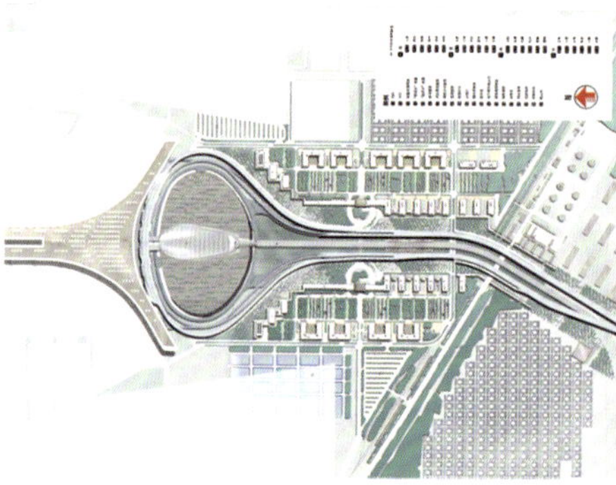
北京首都机场总平面图

城市绿地系统规划
城市各类公园设计
各级自然保护区
风景名胜区
森林公园的规划设计
城市道路广场景观设计
居住区环境设计
各类园区的详细规划和环境设计
城市各类景观体系规划
研究探索项目

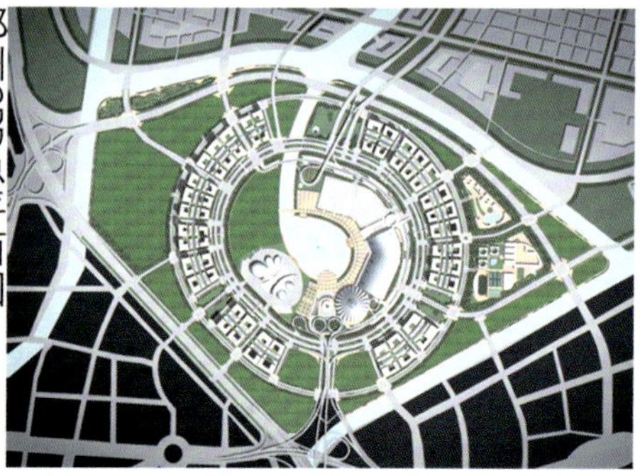
郑州CBD总平面图

四川大学现场实景

Beilin landscape architecture institute' s former body is Beijing forestry university landscape architecture institute .It has the first level qualification of construction design from the ministry of construction .(It' s design qualification is construction First 010132-sj).The country development and Revolution Committee awarded it the construction inquiry first qualification (construction inquiry first 2032101001).As a specialty institute we have the ability of planning, design ,inquiry and so on.

Our institute formed at the beginning of 1980' s, when reform and opening of China started We integrate production, teaching and research together Here advantage is also taken of a team of faculty specialists We believe that “Research will promote production ”,and “ production can stimulate study”.

While many design institutes were deepening their system innovation, we took the lead into the market. “ quality is survival”, serving the society, serving the public is not only our credo. It is our mission our very reason for being

Bemuse of the special course of our institute. We have some special characteristics of our own, which are different from other institutes belonging to the government .Central to the insititute' s philosophy is most widely absorbing the advancd theories and design methods from all around the world .as well as best inheriting and developing the are of ancient Chinese Gardening and landscaping that has enduring value .This pringciple is at the foundation of all that we do to develop successful projects at the request of our society.

Now we have trained tech troop. This troop has good staff, areasonable range of specialties, good staff strcture, and most of them are in the prime of their life and brave in design exploring.Having worked over 30 provinces and zhixia cities of china in 20 years, the insititut' s reach is truly national. Several planning design and construction projects have received honors and won competitions. Examples such as “ sanya green space system planning "which won the ministry of construction praise award honor , “Xianhu landscape arboretum design of Shengzhen City “which received the third prize award of the ministry of construction , “lijing Garden apartment environment design” and “New campus environmental landscape design of zhejing university” won the ministry of forestry a first class award. Those featured projects showed that our institute is a competitive one among so many institutes in land-scape architecture, urban design and environ mental planning

北京中国风景园林规划设计研究中心

BEIJING CHINA RESEARCH CENTRE OF LANDSCAPE ARCHITECTURAL DESIGN AND PLANNING

北京中国风景园林规划设计研究中心成立于1993年，隶属于建设部中国风景园林学会，是国内从事历史与自然保护区、风景旅游区、城市园林和城市公共景观研究设计的专业设计公司，具有甲级设计资质。

从1996年起至2005年7月，我中心已经承担和完成约150余项各类景观规划设计任务，其中仅大型园林设计项目就有40多项，城市公共景观项目70多项。在长期的研究和设计实践过程中，我们逐渐形成了以“生态.人文.共生”为核心的设计理念，不断探索创造现代都市景观与传统中国文化相结合的“中而新”的设计风格。综合考察、统筹协调项目内涵和所处环境的生态、生活、自然、技术、文化、休闲、市政、交通、经营、保护、避险、防灾、以及历史沿革与现实需要等等各种景观因素，使之层次多样、丰富完善，同时又能相得益彰、和谐共存，为实现和谐社会、创造宜居城市和宜人环境不懈努力。

北京市韦伯豪住宅区

“充满活力、安定有序、人与自然和谐相处”，这是现代社会发展的目标，也是现代景观建设、景观设计所要创造和达到的境界。北京中国风景园林规划设计研究中心积十余年、百余个项目的成功经验，愿以先进的设计理念、高科技的设计手段，与所有致力于现代景观建设的朋友们一道，为创造我们共同共存的美好环境、建设美好的未来贡献力量。

中国国际展览中心新馆环境设计

北京市赢海名居别墅住宅区

北京市月坛公园

浙江天台山国家级风景区

业务范围：

- 风景名胜区及旅游经济区的规划设计
- 小城镇历史名城、园林城市的规划设计
- 城市绿地系统规划
- 别墅区、度假区规划设计
- 城市园林、绿地、城市广场及城市街景的规划设计
- 景观建筑设计
- 主题公园及室外公共娱乐项目的规划设计
- 特色建筑的室内外景观设计
- 各类住宅区的景观改造及规划设计等

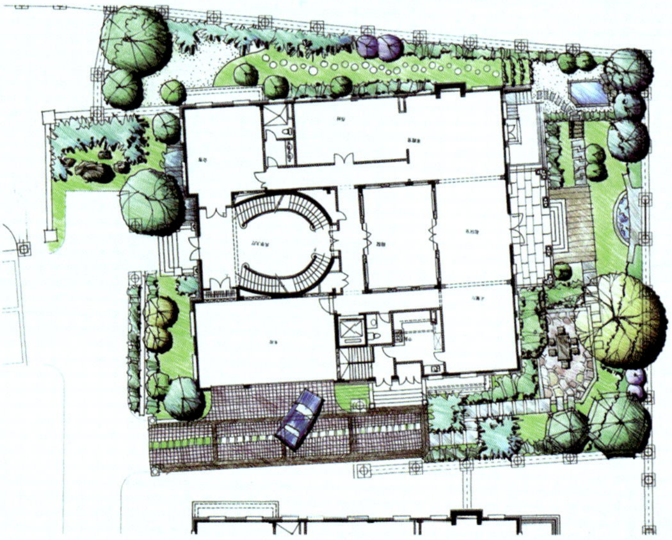

北京市财富公馆别墅居住区

公司地址：中国北京海淀区知春路甲48号盈都大厦5A　邮编：100086

TEL：（010）58732833/32/71　FAX：（010）58732873

网址：WWW.chinagarden.cn